The mathematical science of
Christopher Wren

The mathematical science of Christopher Wren

J. A. BENNETT

Curator of the Whipple Museum of the History of Science,
University of Cambridge.

CAMBRIDGE UNIVERSITY PRESS

CAMBRIDGE
LONDON NEW YORK NEW ROCHELLE
MELBOURNE SYDNEY

PUBLISHED BY THE PRESS SYNDICATE OF THE UNIVERSITY OF CAMBRIDGE
The Pitt Building, Trumpington Street, Cambridge, United Kingdom

CAMBRIDGE UNIVERSITY PRESS
The Edinburgh Building, Cambridge CB2 2RU, UK
40 West 20th Street, New York NY 10011–4211, USA
477 Williamstown Road, Port Melbourne, VIC 3207, Australia
Ruiz de Alarcón 13, 28014 Madrid, Spain
Dock House, The Waterfront, Cape Town 8001, South Africa

http://www.cambridge.org

First published 1982
First paperback edition 2002

A catalogue record for this book is available from the British Library

Library of Congress catalogue card number: 81-18045

ISBN 0 521 24608 3 hardback
ISBN 0 521 52472 5 paperback

To My Mother and Father

Contents

Preface

It is natural that all of us first hear about Christopher Wren through his most obvious legacy: the churches, the palaces and hospitals, and the cathedral, that have graced London for three centuries. The buildings, however, are the products of Wren's intellectual maturity and until the age of thirty-seven, he was professionally engaged in astronomy while maintaining broad interests in other branches of mathematics and natural philosophy. The imbalance in the basis of Wren's popular acclaim is understandable, but the scholarly neglect of an important source for appreciating his attitudes and philosophy is not. This book is an attempt to amend the neglect and to approach Wren by following the route of his own intellectual development.

It was Michael Hoskin who first suggested to me that Wren's science would be a fruitful topic for research, and much of the basic work was done while preparing a Ph.D. dissertation at Cambridge under his supervision. I am grateful to him and to Alistair Crombie for encouraging me to develop the ideas and material further into a book. I want especially to thank my wife France for her help and encouragement while I was preparing the dissertation and writing the book.

J.A.B.

1

Introduction

So much has been written on Wren, and so much research dedicated to his memory, that a new and different approach must begin with a little historiography to explain how it will complement what is already known and what particular problems already exist that it might help to solve.

That it is misleading to divorce the history of ideas from the study of their social, cultural or political contexts is now commonplace. The same is true of divisions within intellectual history – between the histories of science and art, technology, philosophy or economics. Organizational divisions, necessary in their own way, are difficult to bridge, not least because it is rarely clear where we should begin to build. The distortions they create are more commonly brought to light in the kind of masterly synthesis that establishes a new approach to an area of history, than in a scholarly monograph on a restricted and clearly defined subject.

Yet, occasionally, a problem can arise, whose source in an artificial division of labour is not difficult to guess. Just such a case is Wren.

From a modern point of view, Wren divided his career between two subject areas – science and architecture. As we look at the seventeenth century, the subjects seem even more distinct than today, since the decorative element played a much greater role in architecture. The fact that Wren now falls within the histories of both science and art has created many questions surrounding the change he made in the mid-1660s. He was educated, trained and enthusiastically involved in natural philosophy. At the height of a brilliant career he apparently abandoned science for what proved to be an even more successful career in architecture. How was he expected to fill the important post of Surveyor-General at a time of acute crisis? Of what relevance was his previous experience? How did he make so fundamental a shift?

We will find that, looked at within a seventeenth-century context, this change becomes less significant than it appears at first, but it leads on to more important questions. Wren's science has been persistently neglected by his biographers and by the art historical studies of his architecture. It is natural, of course, to begin to look at Wren through his buildings, his most obvious legacy. But to neglect his early work and the period of his intellectual development is to lose a vital source in understanding his mature work in architecture. As we shall see, it is misconceived to couch this problem in terms of how Wren's 'science' influenced his 'architecture', but a study of his natural philosophy ought not only to fill a gap in our knowledge of his life, but also to enhance our understanding of his work as a whole.

Wren's science has been neglected. There are articles on particular subjects and isolated problems, and each of the many books recognizes his early career as an astronomer and his general interest in many branches of natural philosophy. What we seem to need is a fuller, coordinated account that can aim at general interpretative conclusions and, in turn, at analyses of particular problems such as our first – what one writer has called the 'scientist, suddenly turned architect' (Fuerst, 1956, p. 119).

The second problem, that of the influence of Wren's early career, is especially acute, since the one-sided approach to Wren has created difficulties – particularly when it comes to interpreting the theory behind the architectural form and relating it to an intellectual context. To understand this we begin by following a few of the more sensitive and penetrative studies of Wren, in their progress backwards from the buildings to the philosophy and its source.

The art historian typically approaches Wren's work in terms of an established critical framework derived from the stylistic development of contemporary European architecture. The most obvious feature of this framework are the two categories Renaissance and Baroque, and the primary question is how Wren's architecture will relate to the developing European style when England has played only a small part in the mainstream progression. The conclusion, it seems, is that Wren does not fit easily into either camp.

In his classic *Outline of European architecture,* Pevsner sums up this conclusion by saying that: 'Wren's style in churches and palaces is classical, no doubt, but it is a Baroque version of classicism' (Pevsner, 1968, p. 326). Wren seems to fall between the Renaissance and the Baroque, while somehow partaking of both at once. Looked at one way, the restraint and order of Wren's intellectual approach to design would place him on the classical side, and yet he feels free to break established rules and use classical elements in unconventional ways. From another view, this 'impurity' of style gives his work an affinity with the more free expression of the Baroque, and yet he never seems really in sympathy with the true Baroque spirit.

These conclusions are not, of course, original, but are drawn out of the important accounts of Wren's architecture. The tension which is common to each analysis is something more subtle and important than the mere fact that Wren's style changes and develops through his career. Summerson, for example, speaks of 'a strain of inconsistency running through the whole of Wren's work' (Summerson, 1949, p .79). Again, Summerson writes of work that 'falters between the static unity of the "high" Renaissance and the dynamic, emotional unity of the Baroque' (Summerson, 1949, p. 85), and Sekler describes Wren as 'a follower of a classical ideal forced by the artistic climate of his time to achieve Baroque creations' (Sekler, 1956, p. 94).

The language of these analyses is very significant, with key words such as 'inconsistency', 'falters', 'forced'. To accept them is not to conclude that Wren himself was either inconsistent, or faltering, or forced, but that this is how he appears, viewed from a particular critical standpoint. Perhaps Summerson best sums up the position with a nicely apologetic sentence, where he says:

That Wren was not by temperament in accord with the Baroque spirit is perfectly clear; but it is equally clear that in such designs as Hampton Court and Greenwich he was handling classic forms in a loose, unconventional fashion which, allowing for a strong individual trend, can be called by no other name.

Summerson (1949), p. 84.

The conclusion is extremely interesting and perfectly valid, provided we remember that its dualism has not originated from an internal study of Wren, but has been imposed from without, originating in a development not necessarily related to Wren's own thinking. A study of Wren's thought and intellectual background ought to help us to understand why his architecture should appear thus from this viewpoint. However, serious problems do arise when the analysis is carried further and used as a key to understand Wren's own ideas.

Of Wren's five commentaries on the theory of architecture, the one that seems to promise most in this respect is that labelled 'Tract I' by his son and published in the *Parentalia*. [1] One key paragraph has always been quoted as a summary of Wren's philosophy of beauty:

There are natural Causes of Beauty. Beauty is a Harmony of Objects, begetting Pleasure by the Eye. There are two Causes of Beauty, natural and customary. Natural is from Geometry, consisting in Uniformity (that is Equality) and

Proportion. Customary Beauty is begotten by the Use of our Senses to those Objects which are usually pleasing to us for other Causes, as Familiarity or particular Inclination breeds a Love to Things not in themselves lovely. Here lies the great Occasion of Errors; here is tried the Architect's Judgment: but always the true Test is natural or geometrical Beauty.

C. Wren (1750), p. 351.

Now, each writer on Wren has taken 'natural' and 'customary' to correspond with familiar categories in architectural history – natural beauty, 'from Geometry', being that of the Renaissance theorists who held that correct forms and proportions could be demonstrated objectively on geometrical grounds and could provide an unchanging rule-book for design; and customary beauty, 'begotten by the Use of our Senses', being the more subjective criterion of the moderns, a sanction for free and imaginative expression and the architecture of the Baroque. So, it is said, Wren identified two kinds of beauty, and this corresponds neatly with the result of stylistic analysis where, in practice, he seemed to embrace, in part, both the classical and the Baroque.

There are two immediate problems. One is that we have already seen Wren condemn customary beauty as 'the great Occasion of Errors'. He seems rather to introduce it as a warning, and his clear preference (leaving aside for now just what the concept was) was for natural beauty: 'always the true Test is natural or geometrical Beauty'.

Some writers (for example Sekler, 1956, and Whinney, 1971) accept this and they are left with an unsatisfactory divergence between theory and practice for, if Wren's 'natural or geometrical Beauty' corresponds to the Renaissance concept, he was far from consistent in its application.

Others minimize Wren's strictures and link the compromise they have seen in his architecture with a corresponding ambiguity in theory. Downes finds it 'entirely in character that Wren admits two kinds of beauty' (Downes, 1971, pp. 48–9), while Summerson says that: 'Wren joins hands with the most primitive and the most modern exponents of architectural aesthetics' (Summerson 1953, p. 134). The difficulty here, of course, is that the two concepts of beauty are incompatible and it would be difficult to take Wren seriously as an aesthetic theorist if he did not realize this.

Fuerst carries this position to its logical conclusion, when he says that Wren's theory 'reveals a latitudinarianism capable of embracing two entirely different, contradictory and mutually exclusive canons of art' (Fuerst, 1956, p. 173), and concludes that:

Wren's work can show parallels to the Renaissance as well as to the Baroque, and that Wren could regard beauty from the contradictory points of view of classicism and romanticism.

Fuerst (1956), p. 178.

Can we seriously believe that Wren was so oblivious to the implications of his theory? The source of so curious a stance has been sought in the fashionable attitudes of the Royal Society – 'scientific scepticism . . . doubt in immutable standards and *a priori* assumptions' (Fuerst, 1956, p. 175). But the virtuosi will hardly thank the historian who finds their scepticism the source of such muddled thinking.

No, this conclusion itself indicates that, somewhere, we have taken a wrong turning. It must surely be better to begin at the beginning, to examine this scientific background of Wren's – its nature and content – and allow it to throw light on both his aesthetics and his architecture.

An account of Wren's natural philosophy, as well as revealing, in detail, a neglected side of his life, ought to tackle certain particular problems. One is the nature of the shift Wren made in mid-stream from science to architecture. The other is the attitudes and philosophy he carried with him. A study of his formative years must help with some of the difficulties that have dogged a full understanding of the whole of his work.

2

The mathematical sciences

Mathematicks, (at that time, with us) were scarce looked upon as *Academical*
Studies, but rather *Mechanical;* as the business of *Traders, Merchants,
Seamen, Carpenters, Surveyors of Lands,* or the like, and perhaps some
Almanack-makers in London . . . For the Study of *Mathematicks* was at that
time more cultivated in *London* than in the Universities.

Hearne (1725), vol. 1, pp. 147–8.

Thus wrote John Wallis, Wren's fellow mathematical professor at Oxford,
of his own introduction to mathematics at Cambridge around 1635. We can
understand why the young Wallis, his undergraduate studies devoted to
what was a mediaeval curriculum, should, by comparison, have seen
contemporary mathematics as practical rather than academic – the
province of merchants, navigators and surveyors. For there was, in
England, a vigorous tradition of practical mathematics, linking mathematicians
and mathematical teachers with the practitioners of the mathematical
sciences – instrument-makers and their clients. While there was a sustained
interest in theoretical developments, and here in time the University men
would play an important role, this interest was an integral part of a subject
whose overall character we would recognize as 'applied'. The exploitation
of mathematics to practical ends, particularly in the use of mathematical
instruments, was a dominant force, and its focus was, appropriately, in
London. Wren himself, in 1657, said that:

I must congratulate this City, that I find in it so general a Relish of
Mathematicks, and the *libera philosophia,* in such a Measure, as is hardly to
be found in the Academies themselves.

C. Wren (1750), p. 206.

We shall see that the character of Wren's work in natural philosophy
places him within the mathematical sciences tradition in England, so we
must first outline the nature of this tradition – how it began and developed,
the subjects it embraced and the philosophy it espoused.

A new and different interest in mathematics in England can be traced to the mid-sixteenth century, when demand for improved techniques of navigation, fortification, surveying and cartography can be related in turn to political and economic conditions.[1] Whatever the reasons for its flowering, the character of the subject, for a century or so, was set at this time by a small group of practitioners, notably Robert Recorde, John Dee, and Leonard and Thomas Digges. It can be summarized in their two fundamental tenets: that mathematics is both certain and useful. Mathematics is indeed the only source of certainty in the natural world, and it can be applied to a particular domain of subjects known as the mathematical sciences. The thesis could be developed and the domain expanded at length, but some of the most important of the mathematical sciences were surveying, navigation and astronomy. This utilitarian philosophy led to an important emphasis on instrumentation and to both close liaison and overlap between the mathematician and the instrument-maker.

Recorde first published his arithmetic textbook, significantly entitled *The grounde of artes,* in 1542, and his geometry textbook, *The pathway to knowledge,* followed in 1551. Already, the characteristics of the tradition were emerging as, in both works, Recorde stresses the certainty and usefulness of mathematics. Thus he who seeks knowledge, 'must also before all other arts, taste of the mathematical sciences' (Recorde, 1551, dedication). In his preface to *The pathway,* 'declaring briefely the commodities of Geometrye, and the necessitye thereof', Recorde establishes his subject as the foundation of all the arts, drawing particular attention to building, and continuing:

It should be to longe and nedelesse also to declare what helpe all other artes Mathematicall haue by geometrie, sith it is the grounde of all theyr certeintie, and no man studious in them is so doubtful therof. For he can not reade ii. lines almoste in any mathematicall science, but he shall espie the nedefulnes of geometrie.[2]

In his more advanced arithmetic, *The whetstone of witte* (1557), he is emphatic:

It is confessed emongeste all men, that knowe what learnyng meaneth, that besides the Mathematicalle artes, there is noe vnfallible knowledge, excepte it bee borowed of them.

Johnson (1937), p. 134.

Recorde's *Castle of knowledge,* his astronomical textbook of 1556, is described by F.R. Johnson as 'the first comprehensive, original treatise on the elements of astronomy to be printed in English' (Johnson, 1937, p. 124), and as 'the outstanding introduction to the science of astronomy published during the sixteenth century'[3]. Its well-known, though very brief, account of the Copernican system appeared a mere thirteen years after *De revolutionibus*. The book marks the beginning of an important English tradition in astronomy: a line of successive astronomers would be both innovative themselves, and aware of contemporary advances elsewhere. The tradition would also conform to other characteristics of the mathematical sciences in England, such as instrumentation and practical application.

Recorde's textbooks provided a grounding for the mathematical practitioner and the skilled artisan. But they did more. They established that works in the genre need not be slavish and dull; they could be original, even daringly so. This feature is best known through Thomas Digges's 'Perfit description of the caelistiall orbes', appended to the 1576 edition of his father Leonard's *Prognostication euerlasting*. Here was an English translation of Book 1 of *De revolutionibus*, accompanied by a startling diagram of the heliocentric universe, where the stars were not limited to an ultimate sphere, but spread out in indefinite space.

In 1556 Leonard Digges had added a practical geometry, his *Tectonicon*, to mathematical textbooks in the vernacular, including descriptions of various instruments, and written for 'Surveyors, Landemeaters, Joyners, Carpenters, and Masons'.[4] Works of this kind, appearing in increasing numbers, were providing a basis for the growth of mathematical sciences. As early as 1559, we find, in William Cunningham's *Cosmographical glasse*, that the pupil, when asked whether he has read geometry and arithmetic, can reply that he has been schooled on Robert Recorde: 'Yes sir I haue redde the ground of Artes, The whestone of wytte, and the path way' (Cunningham, 1559, pp.4–5). The master assures him that a grounding in mathematics is essential 'not only for this studies sake whiche you now haue in hande: but for all other artes (whiche taste of the Mathematicalles) that you shall here after trauell in' (Cunningham, 1559, pp. 4–5).

What were the arts 'whiche taste of the Mathematicalles' – the domain of the mathematical sciences? The tradition was only taking shape and its area of competence would become clear only in time. It was defined in practice by the growing number of similar textbooks and the corresponding success in the practical application of mathematical techniques.

An important attempt to formalize its domain was made by John Dee in his preface to Henry Billingsley's English translation (1570) of Euclid. He named and defined the various mathematical sciences – including perspective, astronomy, music, architecture, navigation – all derived from the fundamental sciences of geometry and arithmetic, and set them within an elaborate and comprehensive scheme.

A less ambitious attempt to outline the applications of geometry was made by Dee's pupil, Thomas Digges, in his preface to *A geometrical practical treatize named Pantometria* of 1591. *Pantometria* was the final appearance of the book promised by Leonard Digges in his *Tectonicon* and described as 'a volume conteynynge the flowers of the Sciences Mathematical, largely applyed to our outwarde practise' (L. Digges, 1556, preface). The preface explains the applications of geometry to architecture, fortification, surveying, ballistics, optics:

how necessarie it [geometry] is to attaine exact knowledge in Astronomie, Musike, Perspectiue, Cosmographie and Nauigation, with many other Sciences and faculties, who so meanely trauaileth therein shall soone finde.

L. Digges & T. Digges (1591), preface.

The mathematical sciences comprised astronomy, surveying, measuring quantities of all kinds, cartography, 'perspective', navigation, fortification, engines and machines and – looking forward to our main concern – architecture. There were others but, in spite of Dee, the definition cannot be sharply drawn, and these are the most important. The term 'perspective' covered both the design of optical instruments and drawing in perspective, which was related to surveying. The mathematical science of architecture had three branches – 'civil' (architecture as we understand it), 'naval' (the design of ships) and 'military' (the design of forts and moles).

The mathematical theorists, the teachers, the instrument-makers, the authors and the practitioners themselves (navigators, surveyors, etc.) – the men who carried on the subject – were not formally organized. Groups of like-minded men formed and dispersed; ideas were exchanged; instruments were copied and designs disseminated; authors were plagiarized. For the most part, the English mathematical tradition flourished in an informal way and outside the Universities.

But, in 1597, the nearest thing to a formal organization was begun with the foundation of Gresham College in London, and it might be said that the tradition was institutionalized in the Gresham statutes. The chairs were to include astronomy and geometry – the first mathematical professorships in England. The professors were to lecture in English as well as in Latin. The astronomy professor was to lecture on navigation and the use of nautical instruments, as well as on astronomy; and the geometry professor would teach arithmetic and practical, as well as theoretical, geometry. Mathematics incorporated practical applications in the mathematical sciences.

In the first half of the seventeenth century, Gresham College, its professors and its lectures, provided a focal point for the mathematical sciences. Many of the important practitioners were associated with informal groups, loosely connected with the College, or one or other of the mathematical professors. The first Professor of Geometry was Henry Briggs. He held the chair till 1620, and it was he who established the College as a centre for research and discussion in the mathematical sciences. His associates included such men as William Gilbert, Edward Wright, Thomas Blundeville and William Oughtred, all of whom made important contributions to the development of science in England. Another feature of Briggs's tenure was that he would introduce novel developments, such as logarithms, into his lectures. Later Gresham professors, including Wren, would lecture on recent developments and, indeed, on their own current research.

Briggs's friend and associate, Edmund Gunter, became Professor of Astronomy in 1619 and held the chair till 1626. Gunter designed mathematical and astronomical instruments, and wrote and lectured on practical mathematics and navigation. He also applied geometrical techniques to the theory of fortification (see Gunter, 1662, 'The general use of the canon and tables of logarithmes', ch. 4). Henry Gellibrand succeeded Gunter in 1626. He too was an associate of Briggs and much concerned with navigation. His successor in 1636, Samuel Foster, maintained the tradition of writing popular but sound works in practical mathematics, particularly related to mathematical instruments.[5] His tenure, though interrupted for several years shortly after it had begun, continued till 1652, when he was succeeded by one of Wren's close associates, Laurence Rooke.

During this time the mathematical sciences were flourishing in England, so much so that any attempt to chronicle their growth here would lead us too far from our main purpose. Gresham College, as an informal centre for the dissemination of mathematical knowledge, provided a focus of interest and a place of resort, but the history of the subject ranges much wider. It was during Foster's tenure that some of Wren's friends and colleagues became part of the Gresham circle, and it was from his introduction to this group in 1645 that John Wallis traced its subsequent history down to the foundation of the Royal Society in 1660.

John Wilkins, Jonathan Goddard and Charles Scarburgh were others who were associated with Foster in the 1640s and intimately with Wren in the late 1640s and the 1650s. The link between Wren and the Gresham tradition was strong, even before he was appointed Professor of Astronomy in 1657. Indeed, one profound influence on the young mathematician was William Oughtred, and he had been a friend of Henry Briggs.

The philosophy of the mathematical sciences too was maintained into the seventeenth century. Geometry and arithmetic were usefully applied to relevant subjects, and mathematical instruments were devised and improved. It seemed to Wallis that he was venturing among: 'Traders, Merchants, Seamen, Carpenters, Surveyors of Lands, or the like.'

Yet we have mentioned some significant mathematicians, such as Oughtred and Foster. The mathematical tradition in England did admit the importance of pure mathematics and, while mathematicians emphasized the need to apply mathematical techniques and were willing to communicate them to craftsmen, they were often proud of their own more esoteric skills. The tradition was not *narrowly* practical.

For one thing, writers very often stressed the importance of understanding the mathematical theory behind instrumental techniques and new inventions, rather than merely being able to use the instruments. Strictly speaking, the mathematical sciences dealt with the mathematics to be applied rather than the actual work of the instrument-maker, though, in practice, this did not prohibit mathematicians from making their own instruments. In his preface to Euclid, though written for the 'Common Artificer . . . that dealeth with Numbers, Rule, & Cumpasse' (Billingsley, 1570, preface by John Dee, sig.A.iii, verso), John Dee was very careful to maintain the distinction between the mathematical sciences as such and their realization in 'materiall stuffe' by craftsmen. Thus architecture, understood as a mathematical science, is not the process of building in brick and stone, but 'the Demonstratiue reason and cause, of the Mechaniciens worke: in Lyne, Plaine, and Solid' (Billingsley, 1570, preface by John Dee, sig. d.iii).

Also, pure mathematics was not despised. Thomas Digges was clearly proud to append to his father's *Pantometria* – that 'Geometrical practical treatize' – 'a Mathematicall discourse of the fiue regular Platonicall Solides'. To the charge that this was 'a meare curiouse trifle, seruing to no use or commoditie', he answered:

I nothing mistrust of such as couet the vnderstanding of matters harde and difficile, desiring the knowledge of things somewhat passing the reache and capacitie of the common sorte, wherein onely the nature of man surmounteth beastlye kinde.

L. Digges & T. Digges (1591), preface.

Digges balances both the pure and the practical when he goes on to say that, for the future:

I shall bee prouoked not onely to publishe the Demonstrations of these and many moe strange and rare Mathematicall Theoremes . . . but also to Imprinte sundry other volumes of my Fathers, which hee long sithence compiled in the English toong, desiring rather with plaine and profitable conclusions to store his native language and benefit his Countreymen, than by publishing in the Latin rare and curious demonstrations, to purchase fame among strangers.

L. Digges & T. Digges (1591), preface.

Mathematics was not narrowly practical. Rather, the accepted domain of mathematics included its practical applications in the various mathematical sciences and, if we take an overall view of the character of the subject in England, it was in the applications that the main emphasis lay.

John Wilkins was an immense influence on Wren. The *Mathematicall magick* of 1648 is clearly reflected in Wren's early work and was written in the mathematical sciences tradition. In a work more scholarly and erudite than those we have already mentioned, Wilkins's message was not so much the application of mathematics – that was already accepted; he was writing rather to convince scholars and gentlemen that the practical mathematical sciences were worthy subjects for intellectual study. He entitled its two books 'Archimedes' and 'Daedalus', 'both these being two of the first authors, that did reduce mathematical principles unto mechanical experiments' (Wilkins, 1802, vol. 2, p. 90).

To describe the character of the subject with a single adjective, it might best be called 'Vitruvian'. A parallel use of the terms 'Platonic' or 'Neoplatonic' has become commonplace in recent years. They describe beliefs in an ultimate mathematical model for the world, or the mathematical explanation of natural phenomena. 'Vitruvian' may be used to describe a tradition where the emphasis is rather on the practical application of mathematics in useful arts and sciences.

'Vitruvian' is appropriate because *De architectura*, while admitting a central role to mathematics, is really a practical textbook; it teaches the application of geometry to civil and domestic architecture. It also deals with others of the mathematical sciences, such as astronomy, dialling, machines, surveying and ordnance; and Vitruvius stipulates geometry, optics and arithmetic as components of the architect's education. Frances Yates has pointed to Vitruvius's influence on Dee's 'Mathematical preface' – an influence which extends to the work of other mathematicians in England.[6]

For Dee, of course, there was indeed a higher mathematical order; a whole view of his mathematical philosophy reveals a convinced Platonist. Yet the 'Mathematical preface' is largely, though not exclusively, concerned with practical application, and certainly a wider study of the mathematical sciences in England shows little use of metaphysical justifications to underpin a firm faith in the certainty and usefulness of mathematics. It is absent even at times when we might expect to find it, such as when Digges introduced his treatment of 'the fiue regular Platonicall Solides':

I haue thought good to adioyne this Treatise of the 5 *Platonicall* bodies, meaninge not to discourse of their secrete of mysticall appliances to the Elementall regions and frame of Coelestiall Spheres, as thinges remote and farre distant from the Methode, nature and certaintie of Geometricall demonstration, only heere I intende Mathematically to conferre the Superficiall and Solide capacities of these Regular bodies with their Circumscribing or inscribed spheres or Solides.

L. Digges & T. Digges (1591), p. 97.

The major concern of the tradition was not that mathematics was the key to unlock the mysteries of the universe, but that mathematics was useful in its practical applications. Its philosophy was Vitruvian rather than Platonic.

A biographical study of Wren's early career will show how he came into close contact with this mathematical tradition. We will then be able to recognize the impact it had on his natural philosophy.

3

'That prodigious young scholar'

Wren's limited formal education has been divined and conjectured by his many biographers, but for our purpose it is best simply to consider the known relevant influences. These begin with members of his immediate family: his father, Dean Christopher Wren, and his brother-in-law, William Holder. His father's influence came while the family was at Windsor and was supplemented by Holder's after the Civil War had forced the Dean's retirement to Holder's rectory at Bletchingdon, near Oxford, in 1646. For some years, as the Dean's son became gradually more involved with the life of the University, Bletchingdon remained, in Aubrey's words, his 'home, and retiring-place' (Clark, 1898, vol. 1, p. 403). Its society must indeed have become more and more a secure, and at the same time outmoded, element in the intellectual life of a young man caught up in new and exciting currents of thought.

Wren's son, Christopher Wren, Jnr, whose *Parentalia* is the fundamental source on Wren, was to write in 1740: 'My Grandfather was a Learned Man, skillful in all the Branches of Mathematicks' (C. Wren Jnr to John Ward, 11 August 1740, British Library MS Add. 6209, fo. 209), and in *Parentalia* he quoted a number of the Dean's annotations to works of scientific interest (see C. Wren, 1750, pp. 142–8). From the same source, we find that Wren dedicated some of his earliest exercises in science and instrumentation to his father (C. Wren, 1750, p. 182), and in a letter of 1647 refers to what his father had taught him in mechanics.[1] In the same letter he assumes that the Dean is aware of a tract on dialling by Oughtred, appended to the famous *Key of the mathematicks* which had only just been published.

Our only sources on the nature of Dean Wren's interest in natural philosophy are his extensive annotations to copies of Thomas Browne's *Pseudodoxia epidemica* and of Bacon's *Sylva sylvarum*, now in the Bodleian Library.[2] The notes were written at different times – two dates being 1650 and 1656 – and reveal a well-informed and widely-read interest in mechanics and natural philosophy. Subjects range from astronomy and calendar reform to anatomy, medicine and natural history, and the Dean cites first-hand observation and experiment as well as authority. Authors he refers to include Gassendi, Bacon, Copernicus, Brahe, Galilei, Longomontanus, Kepler and Paracelsus, among many others (a detailed account with references of Dean Wren's annotations is in Bennett, 1974, pp. 30–6). It is clear that the Dean's interests were similar to, and sometimes contemporaneous with, those of his son.[3]

We find Dean Wren strongly opposed to the Copernican theory on both physical and scriptural grounds. Arguments on the physical impossibility of a moving Earth are buttressed by, at times fiercely-worded, appeals to the literal truth of the scripture: 'either, God, or Copernicus speaking Contradictions, cannot both speake' Truthe' (Browne, *Pseudodoxia epidemica,* Bodleian Library, shelfmark 0.2.26 Art. Seld., p. 373). Yet he was familiar enough with the Copernican theory which he believed was for Copernicus 'but a Postulate of Art noe Parte of his Creed' (Browne, *Pseudodoxia epidemica,* Bodleian Library, shelfmark 0.2.26 Art. Seld., p. 291). His alternative was not the old Ptolemaic view but, rather, the compromise Tychonic; he wrote that Tycho 'who lived (52) yeares since Copernicus, hath by admirable, & matchless Instruments, & many yeares exact observations, prooved [the Copernican system] to bee noe better y.ᵖ a Dreame'.[4] According to *Parentalia*, in 1648, at the age of sixteen, his son Christopher devised a lunar theory 'secondum Rationes Tyconianas' (C. Wren, 1750, p. 184).

William Holder, later Fellow of the Royal Society, was a practical mathematician who, according to Aubrey, gave Wren 'his first instructions in geometrie and arithmetique' (Clarke, 1898, Vol. 1, p. 403). He wrote books on speech, music and chronology, in each professing experimental and practical aims and stressing the importance and usefulness of mathematics. However, their mathematical content is undistinguished.

From its preface, and from some of Wren's early interests (see C. Wren, 1750, pp. 195, 241), we can say that Holder's *Discourse concerning time,* revised from earlier papers and published in 1694, may well derive from lessons he gave the young Wren. Because of subsequent revision and additions, we cannot use the source in a straightforward way, but it is typical of his other work that Holder writes;

I do not intend to fall upon nice, Philosophical Disquisitions about the Nature of Time, and Curious Questions relating to it: But upon the Use of it, *in Vita communi.*[5]

So it was due to his father and brother-in-law that Wren's interests first inclined towards mechanics and practical mathematics, but neither Dean Wren nor Holder were able to lead him very far. More significantly, Holder had valuable contacts within the scientific community. He had known Seth Ward and Charles Scarburgh while at Cambridge,[6] and later became associated with the group of natural philosophers whose meetings centred around John Wilkins's rooms at Wadham College, Oxford. It was, no doubt, due to Holder that Wren came into contact first with Scarburgh and his circle in London, and then with the Oxford group. It was through these contacts that Wren came to know a broader, more liberal, attitude to science and met a more rigorous context for serious work than the pious dilettantism of his father or the well-meaning amateurism of Holder.

Scarburgh was a Royalist physician with interests in mathematics, navigation and astronomy, and important contacts among mathematical practitioners.[7] It was he who introduced Wren to the work of William Oughtred (see C. Wren, 1750, pp. 185–6). According to Wallis, Scarburgh had been a member of the group meeting at Samuel Foster's rooms in Gresham College around 1645.[8] It seems unlikely that he joined in their meetings as early as this (see McKie, 1960, p. 15), but that he was associated in Wallis's mind with the Gresham circle is consistent with his interests. By about 1647, when Wren was staying with Scarburgh, probably first as a patient but later as a guest and pupil, his London home was a meeting place for natural philosophers, especially those of Royalist persuasion (see Pope, 1697, pp. 18–19, 117–18). Aubrey says that it was Scarburgh and Holder who suggested to John Greaves that Seth Ward should succeed him as Savilian Professor at Oxford.[9]

So Wren spent some time, as he put it, 'greatly enjoying the society of the famous Physician' (Milman, 1908, p. 19), and could write to Oughtred that he owed to Scarburgh 'any little skill that I can boast in Mathematics' (Milman, 1908, p. 21). Scarburgh had encouraged Wren to prepare a Latin translation of a treatise on dialling by Oughtred, to gain both the author's favour and 'that of all those Students of Mathematics who acknowledge Dr. Oughtred as their Father and Teacher' (Milman, 1908, p. 20). Constructing sundials was one of the earliest expressions of Wren's own talent for the mathematical sciences (see Clark, 1898, vol. 1, pp. 403–4).

Clearly, Wren was becoming known to some influential natural philosophers, and this, along with his personal contacts, accounts for his easy acceptance into the scientific community at Oxford after joining Wadham College as a gentleman commoner in 1650 (see Hutchison, 1976, p. 22). It may be that Dean Wren knew John Wilkins, Warden of Wadham;[10]

certainly, Holder was a friend of Seth Ward, whom Wilkins had persuaded to come to Wadham in 1649. In only Wren's first year of residence, no less than the Savilian Professor of Geometry, John Wallis, told Hartlib that Wren was highly commended by Wilkins (see G. H. Turnbull, 1952, p. 108).

At Oxford, Wren met other members of the Gresham circle under Foster, namely Wilkins, Wallis and Goddard; indeed, he came to describe himself as 'a most addicted Client' of Wilkins.[11] He joined in the activities of the group that met in Wilkins's rooms during the 1650s. Only scattered references to Wren's work survive, largely in chance remarks recorded by others, but he was clearly one of the most active and enthusiastic members. An account of the group's work, written in a letter from Wren to William Petty in 1656, is one of our few sources on their activity (see Bennett, 1973, pp. 146–7), and *Parentalia* contains a list of fifty-three entries entitled:

New Theories, Inventions, Experiments, and Mechanick Improvements, exhibited by Mr. Wren, at the first Assemblies at Wadham-College in Oxford, for Advancement of Natural and Experimental Knowledge, called then the New Philosophy.[12]

By drawing together some of these scattered remarks, we can see where Wren's interests were leading. They were wide-ranging, but the predominant themes were mathematics, astronomy and instrumentation.

In 1650, Wallis told Hartlib that Wren, 'besides divers other fine inventions and contrivances, hath found out a way to measure the moistnesse and dryness of the air exactly'.[13] Wren himself visited Hartlib in 1653 and, most significantly, the same year, Ralph Greatorex told Hartlib of Wren's perspectograph (see G. H. Turnbull, 1952, pp. 114–15). Greatorex was an instrument-maker, formerly apprenticed to Elias Allen and, like Allen, a friend of Oughtred. Among his customers were Ward, Wilkins and Goddard; and another of his associates and a fellow instrument-maker, Christopher Brookes, became Manciple of Wadham College under Wilkins. Brookes was Oughtred's son-in-law (see E. G. R. Taylor, 1954, p. 234). Wren was clearly well-placed to strengthen his links with the English practical mathematical tradition.

In the same context, an important venture for the mathematical circle at Oxford was a Latin edition of Oughtred's *Key of the mathematicks (Clavis mathematicae)* published there in 1652. The bulk of the translation was by Robert Wood, except that the appendix on dialling had already been translated by Wren at Scarburgh's suggestion. In the preface, Oughtred acknowledges the parts played by Wallis, Wood, Scarburgh, Ward and Wren:

a youth generally admired for his talents, who, when not yet sixteen years old, enriched astronomy, gnomics, statics and mechanics, with brilliant inventions, and from that time has continued to enrich them, and in truth is one from whom I can, not vainly, look for great things.[14]

Oughtred presented Wren with an inscribed copy (British Library MS Add. 25 071, fos. 34, verso, and 86). The passage has often been quoted in laudatory accounts of Wren's life, but its real significance is as a public testimonial from the father-figure of the tradition, that here was a rising star in the mathematical sciences.

A similar group of names, representing the mathematical interest of the Oxford group, occurs in a reference of about 1652, when Wallis says he presented problems relating to the quadrature of the circle to Ward, Rooke, Richard Rawlinson, Wood and Wren (see Wallis, 1657, *Operum mathematicorum*, dedication (addressed to Oughtred)). Rawlinson and Wren assisted Wallis with observing a solar eclipse in 1654 (see Wallis, 1655, p. 2). Seth Ward, of course, was the professional astronomer of the group. He set about re-establishing the reputation of the astronomy chair and teaching Keplerian planetary theory. At the same time, he was setting up an observatory at Wadham and 'procureing & fitting Telescopes and other instruments for observation' (Robinson, 1949, p. 70). Wren probably used this observatory, which is generally assumed to have been in the tower over the front entrance (see Van Helden, 1968, p. 215), and may have succeeded to it on Ward's retirement since, in 1663, he paid rent for 'the Chamber over the gate', though by then he had been a Fellow of All Souls for ten years.[15]

John Wilkins was the greatest influence on the young Wren during the early 1650s. In 1655 they were reported to be working together on an 80-foot telescope, 'to see at once the whole moon'.[16] Around 1652 Wilkins was encouraging such interests of Wren's as a double-writing instrument and, already, illustrations of microscopical observations (see British Library MS Add. 25 071, fo. 36). The same year, Wren accompanied Wilkins on a visit to Elias Ashmole (see Josten, 1966, vol. 2, p. 615). The Vitruvian emphasis on practical mechanics and instrumentation of the *Mathematical magick* was manifest also at Oxford, and struck an already established chord with Wren. When John Evelyn visited 'that most obliging & universaly Curious Dr. Wilkins's, at Waddum', he saw a great variety of 'artificial, mathematical, Magical curiosities'. These included:

Shadows, Dyals, Perspectives . . . A Way-Wiser, a Thermometer; a monstrous Magnes, Conic & other Sections, a Balance on a demie Circle, most of them his owne & that prodigious young Scholar, Mr. Chr: Wren.[17]

Wilkins's intellectual influence on Wren was also profound. He had vigorously defended the Copernican theory from the kind of objections raised by the Dean, and could show Wren an altogether more open and balanced approach to natural philosophy than he had experienced before.

The details of Wren's work we can leave to later chapters. His introduction to Oxford was significant in widening his horizons, involving him in an enthusiastic scientific community and establishing many of the important contacts of his career in natural philosophy.[18] Apart from Wilkins and Ward, Goddard would later be a colleague and co-worker at Gresham, as would Rooke who was a close friend of Ward's and of whom Aubrey wrote: 'I heard him reade at Gresham College on the sixth chapter of *Clavis mathematica,* an excellent lecture' (Clark, 1898, vol. 2, p. 204). He too was at Wadham, but his closest links with Wren were in London, first at Gresham and then during the early years of the Royal Society. Wren's closest association with Wallis occurred after Wren's Savilian appointment of 1661, and the same was true of his links with Thomas Willis and Ralph Bathurst. However, they were all at least acquaintances of this earlier period.

Robert Hooke became involved with the group in about 1655 and, like Wren, was encouraged by Wilkins and Ward. He was certainly one of the closest associates of Wren's career, both in natural philosophy and in architecture, and the extent of their collaboration is such that it is frequently impossible, and often inappropriate, to separate their contributions. Robert Boyle came to Oxford in 1656, at Wilkins's invitation (see Warton, 1761, p. 162–3), and soon began to work with Wren.[19] They established a free and wide-ranging collaboration that lasted for about ten years.[20]

Ward's *Vindiciae academiarum* of 1654 is now well known as propaganda for his and Wilkins's version of the new philosophy, but it does at least square with the impression gained from Wren's period in the Oxford circle. Wilkins says in his preface:

there is not to be wished a more generall liberty in point of judgment or debate, then what is here allowed. So that there is scarce any hypothesis, which hath been formerly or lately entertained by Judicious men, and seems to have in it any clearnesse or consistency, but hath here its strenuous Assertours, as the Atomicall and Magneticall in Philosophy the Copernican in Astronomy &c.

S. Ward (1654), p. 2.

From about 1654, Wren's interests, though still extensive, began to move more firmly towards astronomy. We have seen that in that year he observed a solar eclipse with Wallis and Rawlinson, and that in the following year he was working on a lunar telescope with Wilkins. Also, in 1655, Wren told Hartlib of his work on an accurate lunar survey and a theory of the Moon's libration (see G. H.Turnbull, 1952, p. 114). It was about 1654 that Wren began work on the pressing problem of the nature of Saturn,[21] under an influence that would become extremely important, that of Sir Paul Neile.[22]

Neile's son, William, had entered Wadham in 1652 and he proved to be a promising mathematician.[23] Neile himself was associated with Ward[24] and his major astronomical interest was in grinding object glasses of long focal lengths. Improving the telescope was a crucial step in the solution of the Saturn problem, and when Wren revealed his theory in 1658, he said that he did so 'lest the stars would seem to have granted to us the friendship of that very distinguished man, Sir Paul Neile, in vain' (Van Helden, 1968, p. 221).

Although based in Oxford, Wren was not there exclusively during the early 1650s. We know he visited London on several occasions,[25] as did Ward and Wilkins and, in August 1656, Wallis told Huygens that, although Wren was normally in Oxford, he was then temporarily in London (see Huygens, 1888, vol. 1, p. 481). There were more contacts between the two centres than is generally allowed. Wren may have visited Neile's home at White Waltham in Berkshire, since he and Neile were collaborating in 1655 and 1656.[26] Neile had built a workshop and an observatory, and here he liked to entertain his 'chosen astronomical friends'. We know Wren was there in December 1657.[27] Wren was also collaborating over Saturn with William Ball in 1655,[28] though where is not clear.

The appointments of Rooke and Goddard to chairs at Gresham, illustrate again the links between London and Oxford, and Wren was to follow them in 1657. Circumstantial evidence points to a very interesting and, in view of his Royalist background, perhaps surprising conclusion about Wren's appointment.

In 1657, Daniel Whistler married and consequently resigned as Professor of Geometry at Gresham College. However, his successor was already being considered in 1656, and it seems that an appointment was almost made in May of that year. A letter from Cromwell, dated 9 May 1656 and addressed 'For Our worthy Friends the Committee of the City of London for Gresham College', reads:

> We understanding that you have appointed an election this afternoon of a Geometry Professor in Gresham College, We desire you suspend the same for some time, till We shall have an opportunity to speak with some of you in order to that business.

T. Carlyle (1904), vol. 2, p. 493.

Wilkins was in Cromwell's favour and in 1656 he married Cromwell's sister Robina (For Wilkins's links with Cromwell, see Shapiro, 1969). We know he was at Whitehall with Neile on 8 May since, on that day, John Evelyn 'went to visite Dr. Wilkins at Whitehall, where I first met with Sir P: Neale famous for his optic-glasses' (E. S. de Beer, 1955, vol. 3, p. 172). Boyle, who was in contact with Wilkins about this time (see Boyle to Bathurst, 14 April 1656, in Warton, 1761, pp. 162–3), told Hartlib, just after 20 May, that Wren was likely to succeed Whistler as Professor of Geometry (see G. H. Turnbull, 1952, pp. 111–12). Other evidence confirms that Wilkins took a personal interest in appointments at Gresham.[29]

Wren was already known to others who were closely concerned with the succession. Goddard had been appointed Gresham Professor of Physic in 1655, and he too was close to Cromwell. Rooke had left Wadham for the astronomy chair in 1652. Through Goddard or Scarburgh, Wren may even have been known to Whistler since all three were active in the College of Physicians during the 1650s.

It seems, then, that Wren was appointed after the personal intervention of Cromwell, who was acting on the advice of Wilkins, and possibly of Goddard. The part played by the other professors seems to be confirmed by the fact that, while it was the geometry chair that became vacant, Rooke moved from astronomy to geometry, and Wren was inaugurated as Gresham Professor of Astronomy on 7 August 1657 (see J. Ward, 1740, p. 96). In the English draft for his inaugural address, after apologizing for his youth, Wren continues:

I must confess I had never design'd any Thing further than to exercise my
Radius in private Dust, unless those had inveigh'd against my Sloth and
Remissness, with continual but friendly Exortations, whom I may account the
great Ornaments of Learning and our Nation, whom to obey is with me sacred,
and who, with the Suffrages of the worthy Senators of this honourable City,
had thrust me into the publick Sand.

C. Wren (1750), p. 200.

We know the identities of at least some of 'the great Ornaments of Learning and our Nation'. In the circumstance of Wren's appointment it is interesting, and of course significant in Commonwealth England, that Wren also writes that with the advent of the Copernican hypothesis 'began the first happy Appearance of Liberty to Philosophy, oppress'd by the Tyranny of the Greek and Roman Monarchies' (C. Wren, 1750, p. 204). In more than one way he was no longer restricted by the influence of his father.

At Gresham College Wren was, of course, at the historic centre of the mathematical sciences in England, and we have seen from the experience he already had of the tradition that, in spite of his youth, he was well prepared. He could write in his inaugural address that the College had been endowed

hitherto with Men of most eminent Abilities, especially in mathematical Sciences; among whom the Names of Gunter. Brerewood, Gillibrand, Foster, are fresh in the Mouths of all Mathematicians.

C. Wren (1750), pp. 205–6.

Wren also referred to his predecessors' work on logarithms, 'wholly a British Art', and said that 'the whole Doctrine of Magneticks, as it was of English Birth, so by the Professors of this Place was augmented' (C. Wren, 1750, p. 206). As we would expect, Wren was well aware of Gresham's past, and he and Rooke maintained the College's traditional role through their lectures, their research and by fostering group activity and discussion.

Indeed, Wren's Gresham professorship represents perhaps his most active and fruitful period as a natural philosopher. He did not sever his links with Oxford; he kept his fellowship at All Souls and the evidence indicates that he continued to spend a fair proportion of his time there.[30] But London was the new influence, and furnished Wren with further important associations.

His main associates were Neile and Rooke, as well as William Neile, Ball, William Croone, Goddard, Scarburgh and William Brouncker. In December 1657, Wren was making observations at White Waltham, and it was here that, with Neile's assistance (see Wren to Neile, 1 October 1661, Royal Society MS EL.W.3 no. 2) he developed his Saturn hypothesis. Wren wrote of Neile:

This is the man who, having hired the best workmen, ordered the making of these above mentioned celestial devices [telescopes of between 6 and 35 feet], and even greater ones, of 50 feet, in his own house, he himself supervising the work (by virtue of the remarkable strength of his judgment in mathematics). And not less sincerely does he rejoice to share his hospitality at the same place with his chosen astronomical friends; and I am also grateful for the gift of certain remarkable lenses and very many observations of Saturn.

Van Helden (1968), p. 221.

In London, no less than at Oxford, Wren's work was characterized by collaborations and group discussions. A group of 'mathematical friends', meeting regularly in London in 1658, included John Pell, Brouncker, the instrument-maker Anthony Thompson, Neile, Goddard, Scarburgh, Rooke and Wren (see Vaughan, 1838, vol. 2, pp. 478–9). Common interests here centred around mathematics and practical astronomy, so that this was a group likely to cultivate the 'mathematical sciences' tradition. Its historical links too are consistent with this. Pell was a direct link with Briggs and Gellibrand; and Thompson, who was employed also by Wren,[31] had been instrument-maker to Foster (see E. G. R. Taylor, 1954, pp. 81, 220).

With Goddard, Rooke and Wren in residence, Gresham College continued its traditional role as a centre for research. Wren seems to have worked closely with his 'honored freind' (see Royal Society MS EL.W.3 no. 2) Lawrence Rooke and, in particular, they collaborated on experiments to establish the laws of motion, probably in about 1660 (see Chapter 7). Both Rooke and Wren were making astronomical observations: Rooke was especially interested in Jupiter, Wren in Saturn. Goddard was concerned with grinding lenses, and in 1658 Neile and Wren mounted a 35-ft telescope in the College grounds (see Wren to Neile, 1 October 1661, Royal Society MS EL.W.3 no. 2). Thomas Sprat referred to this telescope when in 1659 he wrote to Wren of the condition of Gresham College, then being used as a garrison for soldiers, 'if you should now come to make Use of your Tube, it would be like Dives looking out of Hell into Heaven' (C. Wren, 1750, p. 254).

It was also in 1658 that Wren made his name among European mathematicians and we have already seen that pure mathematics had traditionally been the province of the leading practitioners of the mathematical sciences. He rectified the cycloid, in response to problems set by Pascal; found a geometrical solution to Kepler's problem; and published his solution to the 'Jean de Montfert' problem. It was in 1657 that William Neile had rectified the semicubical parabola, and Wallis wrote that this

was presently seconded with other Demonstrations of the same thing, by Dr. Christopher Wren, . . . , the Lord Viscount Brouncker, my self, and (as I remember,) some others of that meeting, then held at Gresham College, which gave rise to (what is now called) the Royal Society.[32]

Wallis later added Rooke's name to this list (see Huygens, 1888, vol. 7, p. 307).

It is well known that the Gresham lectures of both Rooke and Wren were focal points for meetings of natural philosophers in London. The subjects of Wren's lectures included Saturn, dioptrics and, following the example of Seth Ward, Keplerian astronomy. (For the Keplerian astronomy lectures, see C. Wren, 1750, p. 239.) The statutes required him also to teach navigation and the use of navigational instruments.[33] His lectures of 1659 on the telescope attracted special interest, probably because they related to Wren's own research; Robert Boyle, Robert Moray and Paul Neile were keen to procure copies.[34] We know that Wren presented the results of his study of Saturn in a Gresham lecture, for when the Royal Society requested an account of his hypothesis in 1661, he excused his reluctance on the grounds that

divers there, had been at the trouble to heare the Astronomy Reader at Gresham give fuller discourses on the same subject, w:ʰ he thought then was publication enough.³⁵

It is worth remembering that, on the one occasion when their names were recorded, 28 November 1660, those who 'following according to the usuall Custome of most of them, Mett together at Gresham Colledge to heare M: Wren's Lecture', included no less than Brouncker, Boyle, Moray, Neile, Wilkins, Goddard, William Petty, Ball, Rooke and Abraham Hill (see McKie, 1960, p. 31).

The Restoration brought temporary eclipse for both Wilkins and Ward, but for Wren the loyalties of his father and his uncle Matthew had linked the family name with the Royalist cause. He would now need a sponsor other than Wilkins and it seems that this role was played, at least in part, by Neile. Hartlib wrote to John Worthington in October 1660:

His Majesty was lately, in an evening, at Gresham Coll., where he was entertained with the admirable long Tube, wth which he viewed the heavens, to his very great satisfaction, insomuch that he commanded Sr P. Neale to cause the like to be made . . . for the use of Whitehall. Sr Paul hath very highly commended Mr. Wren to greater preferment, and there is no question but he will find the real effects of it.³⁶

By April 1661 Hartlib could report: 'I have heard great talk of Mr. Wren, and we see frequent changes of preferment' (Crossley, 1847, p. 305). In fact, Wren had been chosen to succeed Ward as Savilian Professor of Astronomy at Oxford early in February and, having resigned as Gresham Professor in March, was installed at Oxford on 15 May.³⁷

As happened with his Gresham appointment, there was apparently no very abrupt change in Wren's pattern of life. Though more generally at Oxford, associating with Boyle and Wallis, and also Millington, Willis and Lower, Wren was often in London during the early years of the 1660s. It is interesting to note that, just prior to Wren's inauguration at Oxford, Christiaan Huygens visited London and on 23 April entertained a group of natural philosophers whose names form a very familiar collection: 's'assemblerent chez moy M. Morre [Moray]. Mil. Brouncker, S: P. Neal, Dr. Wallis, M. Roock, M. Wren, D. Godart' (Huygens, 1888, vol. 22, p. 573). The discussion centred around the theory of impact and grinding lenses.

Sir John Summerson has called Wren's years from 1661 to 1665 'The period of transition from science to architecture' (Summerson, 1960, p. 102), and we will examine this transition in a later chapter. What we have seen of Wren's background and training in natural philosophy is that, in terms of associates, groups and institutions, he was well grounded in the mathematical sciences and likely to be versed in the characteristic features and philosophy of the subject as practised in England. We can expect to find this further emphasized when we turn from the context of Wren's work to consider its content.

In conclusion, it is interesting to note the impression of a visitor to All Souls in the early 1660s. Balthazar de Monconys called on Wren in 1663 and caught his characteristic combination of personal openness and reticence over publication:

i'y allay encore plus pour voir M. Renes grand Mathematicien quoy que petit de corps, mais des plus ciuils & des plus ouuerts que i'aye trouuez en Angleterre: car quoy qu'il ne veüille pas que ses pensées soient diuulguées.

Monconys (1665), vol. 2, p. 53.

4

Astronomy

When Wren began his professional career in astronomy, a hundred years had passed since the Copernican theory was first tentatively expressed in English through Robert Recorde's *Castle of knowledge*,[1] published in 1556. Recorde had also represented the beginnings of the wider domain of the mathematical sciences, so that the 'new astronomy' in England began and developed within this tradition.

It had been a crucial period for astronomy throughout Europe too. If the Copernican theory had not exactly triumphed by the mid-seventeenth century, the Ptolemaic had certainly been replaced as the conservative alternative by the Tychonic where Sun and Moon revolved around a stationary Earth and the planets revolved around the Sun; or by the so-called semi-Tychonic system which accepted a rotating Earth but retained the annual orbit of the Sun. But for astronomers in the vanguard of contemporary debate, a change had occurred more fundamental than the mere juxtaposition of planets in a geometrical problem. With Kepler's *Astronomia nova* a whole new concept of the nature of astronomy had emerged – astronomy, not only as mathematical theory but also as 'celestial physics'. For Kepler, a mathematical model had to be interpreted in terms of a physical mechanism. Though for him the model, of itself, still had an important *explanatory* function, his work led to the notion of planetary theory, rendered intelligible, not through its underlying geometrical principles as it had been from Eudoxus to Copernicus, but through a causal hypothesis.

If we look at how the subject had developed during this period in England, we find several important features. One was that a vigorous tradition in the 'new astronomy' had been quickly established and important developments, both theoretical and instrumental, flowed from it. We shall find that Wren drew on these developments, indeed that he inherited a 'working tradition' in the subject. Second, astronomy had developed as part of the mathematical sciences and partook of many of its characteristic attitudes. Third, it had been in England that Kepler's vision of the nature of astronomy was most enthusiastically accepted – both the new geometrical model with its elliptical planetary orbits, and the physical theory that underpinned it. In espousing the elliptical astronomy, Wren was following the example of a number of English astronomers – Thomas Harriot, William Crabtree, Jeremiah Horrox, Vincent Wing, Samuel Foster, Jeremy Shakerley and, of course, Seth Ward. The notion of 'celestial physics' was perhaps most easily accepted in England because Kepler's causal agent had been taken from a recent English precedent. As Kepler himself wrote:

I have built all Astronomy on the Copernican Hypothesis of the World; the Observations of Tycho Brahe; and the Magnetic Philosophy of the Englishman William Gilbert.

Boas (1962), p. 301.

In the seventeenth century England maintained, in spite of the growing influence of Cartesianism, a tradition of speculation on the role of magnetic or quasi-magnetic forces in the cosmos, a tradition that was an important element in the background to Newton's *Principia*.

Wren's inaugural address (1657) as Professor of Astronomy at Gresham College, is an important and under-used source on contemporary English astronomy. In it, Wren, as well as explaining the virtues and usefulness of astronomy, also depicts its contemporary situation (see C. Wren, 1750, pp. 204–6). He describes the reception of the Copernican theory, the roles of Gilbert – 'Father of the new Philosophy' – and of Descartes, Galileo, and Kepler – 'the *Eudoxus* of this Age, the Inventor of the elliptical Hypothesis'. He runs enthusiastically through the solar system as it now appears in the telescope, referring in passing to Saturn's satellite, announced only the previous year by Huygens. He pays tribute to the recent contributions of his countrymen, and lets slip a few more personal remarks that reflect his own immediate interests:

of all the Arguments which the Learned of this inquisitive Age have busy'd themselves with, the Perfection of these two, Diotricks, and the Elliptical Astronomy, seem most worth our Enquiry

C. Wren (1750), p. 204.

and later,

So large a Field of Philosophy is the very Contemplation of the *Phases* of the coelestial Bodies, that a true Description of the Body of *Saturn* only, were enough for the Life of one Astronomer.

C. Wren (1750), p. 205.

With reference to a solution to the problem of longitude, he remarks that 'former Industry hath hardly left any Thing more glorious to be aim'd at in Art' (C. Wren, 1750, p. 206).

Wren's work on cosmology and on the problem of longitude require chapters of their own, but the mutually related subjects of dioptrics and Saturn are a useful introduction, since they show the intimate connection between theoretical and instrumental advance. This feature appears again with Wren's study of the Moon in relation to the telescope adapted as an instrument of measurement. Indeed, the whole of the present chapter really concerns the development of the telescope and its application to particular problems in astronomy.

To account for the puzzling sequence of telescopic appearances of Saturn was a problem that began with the observations, in Wren's words, of: 'The incomparable Galileo, who was the first to direct a telescope to the sky' (Van Helden, 1968, p. 219). But to devise a model that would explain all the observations that had been published since, was a hopeless task

because observers did not often use very long tubes and absolutely perfect lenses . . . and did not take good enough care to remove completely all superfluous light fringes from the aperture in the customary manner, or because they were unaccustomed to depict graphically on the spot just what they saw distinctly.[2]

A solution could be based only on a series of reliable observations, and became possible only, as Wren said, 'as the mathematicians improved the theory of dioptrics and craftsmen daily promoted the art of working big lenses' (Van Helden, 1968, p. 220). Indeed, to find a solution became a challenge to telescopic astronomy:

distinguished men of nations everywhere, even now, eagerly apply themselves to the production of longer telescopes. Saturn is proposed as the greatest test of skill. This is the target upon which they aim their artfully strengthened vision and they strive to bind this most deceitful star with the laws of a particular hypothesis.

Van Helden (1968), p. 220.

Wren's own observations began in about 1654 (see Van Helden, 1968, pp. 215, 221) and continued at least until 1659 (see Wallis in Huygens, 1888, vol. 2, p. 358), but thanks to Neile and Ball he had access to a series of first-hand observations going back to 1649.[3] Like so much of Wren's work, the research on Saturn was, in many ways, a group effort. Neile provided the telescopes, Neile and Ball contributed independent observations, and Neile and Rooke were consulted over Wren's eventual hypothesis (see Wren to Neile, 1 October 1661, Royal Society MS EL.W.3 no.2). This hypothesis, devised by the most able theorist of the group and presented, appropriately, to the Gresham audience, represented the culmination of a collaborative project stretching over a number of years.

We can pin-point some stages in its development. As early as January 1654/5 (or possibly 1655/6), Wren was making wax models of Saturn (see Wren to Neile, 1 October 1661, Royal Society MS EL.W.3 no.2), and in 1655 he and Neile observed a satellite, but apparently did not recognize it as such before Huygens had announced his discovery in *De Saturni luna* (1656).[4] In the same year (1655), Ball first saw a dark band on the body of Saturn (the shadow of the rings) and, wrote Wren, 'showed it to us at once'. Wren saw 'a certain zone, darker than the rest of the area of the disc and slightly narrower than Jupiter's belts', and thought he could make out a series of spots (see Van Helden, 1968, pp. 223–4). By 1656 Wren felt sufficiently confident of the group's progress to write to Petty that on the basis of 'many exact Pictures of ♄ not only of his Ansulae but his Spots',

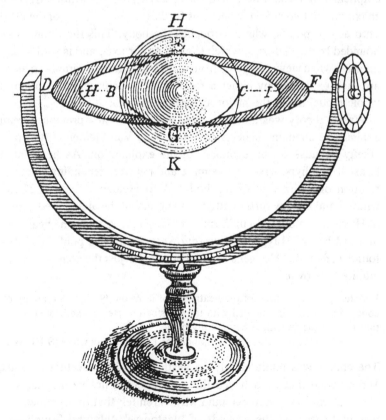

Fig. 1. Wren's model of Saturn, from his *De corpore Saturni* (Reproduced in Huygens, 1888, vol. 3, facing p. 424.)

they had 'attain'd to a Theory of his Rotation & various inclination of his Body' (Bennett, 1973, p. 147). As for the nature of the 'Ansulae' themselves, Wren formulated his hypothesis while observing at White Waltham in December 1657 (see Wren to Neile, 1 October 1661, Royal Society MS EL.W.3 no.2). He made two pasteboard models, lectured on his ideas at Gresham, and:

The hypothesis made more durable in metal was posed on the top of that Obeliske, wch was erected at Gresham Colledge in May 1658 . . . to rayse the 35 foot Telescope.[5]

It was about this time that he wrote *De corpore Saturni,* intended as a preliminary sketch, a 'short and generall account', to be followed, after further observation, by a more complete treatise.[6]

According to Wren's model (see Figure 1), Saturn is surrounded by an elliptical ring (which he called the 'corona') whose width varies from a maximum at two points, where it is furthest from the body of the planet, to zero at two points, where it touches the body. This fluid corona is thus bounded by two ellipses, which share a minor axis, and is so thin as to be invisible when viewed edge-on from the Earth. Saturn's various appearances are then explained by either a rotating or a reciprocating motion of the whole model about the major axis of the corona.

Wren's theory was a considerable advance on any previous attempt to explain the changing aspects of Saturn (see Van Helden, 1968, p. 217), chiefly because of the economy of his explanation. As he pointed out himself, 'the hypothesis is so simple and natural, depending solely on the rotation or inclination of the body' (Van Helden, 1968, p. 222). Yet, without making any further attempt to publish it, he abandoned it in favour of Huygens's model of a uniform symmetrical ring, which he heard of early in 1659 before Huygens's *Systema Saturnium* was published.[7] He had found, in fact, that Huygens's model excelled in just the points where he had admired his own:

I confesse I was so fond of the neatnesse of it, & the Natural Simplicity of the contrivance agreeing soe well with the physicall causes of the heavenly bodies, that I loved the Invention beyond my owne.

Wren to Neile (1 October 1661), Royal Society MS EL.W.3 no.2

The choice was made clearly on aesthetic and intellectual grounds, for Wren added that he believed the models could not be distinguished by observation. We can understand why Wren felt that the symmetrical ring was preferable on the grounds of 'neatnesse', 'Natural Simplicity' and consonance with 'the physicall causes of the heavenly bodies', but when

Huygens read Wren's account early in 1662, he was impressed and said that he was astonished that Wren had not shown it to him when they had met in London the previous year (Huygens, 1888, vol. 4, p. 24).

De corpore Saturni had reached a larger audience only in 1661, after Frenicle de Bessy had sent his theory on Saturn to Kenelm Digby, and Digby had read it to the Royal Society on 4 September (see T. Birch, 1756, vol. 1, p. 41; note also p. 47). Neile and Wren were present and, after Neile had pointed out that Wren had devised a solution similar to Frenicle's some years previously, Wren admitted that this was so, but said that he had abandoned it in favour of Huygens's model (see Huygens, 1888, vol. 3, pp. 368–9; note also p. 425). Digby wanted to have a copy of Wren's hypothesis, and Wren was asked to send one to the Society, so that a copy could be sent via Digby to Frenicle.[8] Wren was reluctant to resurrect his old work and, as a compromise, sent his only copy to Neile, asking him to let it go no further and saying:

I would not proceed with my designe, nor expose soe much as this sheet any farther then to the Eye of my bosome freind to whome even my errors lay alwaies open.[9]

However, without Wren's consent, copies were sent to Frenicle (via Moray and Digby) and to Huygens.[10] Apparently, this was to show Frenicle that his theory had already been suggested and abandoned (see Huygens, 1888, vol. 3, pp. 368, 425), but the move was ill-judged, since it gave a false impression that Wren was opening controversies with both Huygens and Frenicle.[11] Moray at least kept Huygens correctly informed of Wren's real opinions,[12] so that Huygens wrote of Wren's hypothesis:

elle se pouroit mieux defendre et partant quand il la quite pour la miene il montre qu'il aime plus la veritè que ses propres inventions.
Huygens (1888), vol. 3, p. 437.

The episode provides several further points of interest. Wren's apparent reluctance to publish is quite consistent with attitudes he adopted on other occasions. *De corpore Saturni* was only a preliminary sketch; several points required further observations (see Van Helden, 1968, p. 224), and Wren seems to have preferred completeness and certitude before formal publication. He planned a longer treatise, but was typically impatient with any undertaking of the kind; a creative momentum often carried him on to fresh topics and new problems (note Wallis's remarks in Huygens, 1888, vol. 2, p. 358; note also p. 305). That he accepted Huygens's theory and suppressed his own is also quite in character:

to see ingenious men, neglecting what was well determined before, to doe worse on the same subject because they would doe otherwise, was alwaies wont to make me passionate.

Wren to Neile (1 October 1661), Royal Society MS EL.W.3 no.2.

If the imbalance between Wren's reported activity (and thus his scientific reputation) and the record of his finished work seems strange, it is worth remembering the unlikely sequence of events that unearthed and preserved his *De corpore Saturni*. We should remember too that Wren had given his hypothesis publication of a sort: he lectured on his theory at Gresham College, it was discussed there and a model was displayed publicly. That he chose this informal promulgation is consistent with the picture of Gresham College outlined in Chapter 2. We have seen that Wren wrote, with reference to the Royal Society:

divers there, had been at the trouble to heare the Astronomy Reader at Gresham give fuller discourses on the same subject, wch he thought then was publication enough.

Wren to Neile (1 October 1661), Royal Society MS EL.W.3 no. 2.

Wren's interest in Saturn covered the period 1654–9, and we have seen the significance of improved telescopic observation in this context. If we look at his contemporary attempts to improve the telescope, we seem to uncover a carefully considered programme in which practical advances were to be based on a theory of optics, dioptrics and vision. Wren pursued this at Oxford and at Gresham, in association with Neile and, to a lesser extent, with Goddard. As is usual with Wren, the record has been drawn together from scattered references, but the result is a valuable example of what English astronomers were attempting during this period.

In 1656 Wren wrote to Petty:

It is not I suppose new to you what we have been doeing in Dioptricks, which we have improv'd both as to the Theory in giveing a true account of Refraction and of vision, (as that the Chrystalline Humor is not the Principle Instrument of Refraction in the Eye, nor essential to Vision but meerly to convenient Vision.) of Perspectives & Microscopes, in which we give the Reasons and Rules for charging them, or suiting the Eye Glasses, and giveing them their due appertures, things formerly done by Experiment only; and as to the Practick, thô we have not yet arriv'd to a good 50 ft Glass, yet we have very long ones from 12 to 36 ft.

Bennett (1973), pp. 146–7.

Sprat also says in his *History* that Wren derived optimum arrangements of lenses and diaphragms from a mathematical account of refraction.[13] When Sprat claims that 'He discours'd to them a Natural and easie Theory of

Refraction, which exactly answer'd every Experiment', he probably refers to the Gresham meetings which preceded the formal establishment of the Royal Society.[14] We know of the interest aroused by Wren's 1659 lectures on telescopes at Gresham,[15] and he may then have stressed the importance of a theoretical basis, since his son's catalogue of surviving tracts includes 'Lecturae anglicae & latinae, de luce & refractione' (C. Wren, 1750, p. 241). The Gresham professors were expected to lecture in both English and Latin.

One of the sources for Wren's work was certainly Kepler. In his Gresham address of 1657, while Wren described Kepler as 'that great foreign Wit... the Compiler of another new Science, Dioptrics' (C. Wren, 1750, p. 204), he predicted confidently that the current work of the English astronomers would see the science perfected:

the Perfection of both these [dioptrics and elliptical astronomy] are justly to be expected from Men of our own Nation at this Day living, and known to most of this Auditory, the Clarity of these latter, makes me cease from a larger Encomium of Kepler

C. Wren (1750), p. 204.

Both Kepler and Descartes had treated the optical properties of the eye as part of the domain of dioptrics, that is, as part of the optical apparatus, with dioptrics dealing with the behaviour of light between its source and the retina. Since the eye is the 'given' element, dioptrics must begin with the nature of light and the properties of the eye before turning to the means of improving vision.

Wren adopted this domain, giving 'a true account of Refraction and of vision', and typically emphasized the importance of precise measurements. Thus Sprat says, 'He has exactly measur'd and delineated the Spheres of the humors of the Eie, whose proportions one to another were only ghess'd at before' (Sprat, 1667, p. 314). The measurements were taken from the eye of a horse, as Wren recalled in a letter to Brouncker:

I once surveighed an horses eye as exactly as I could, measuring, what ye diameters of ye severall spheres of ye Humors were, and what ye proportions of ye distances of ye Centers of every sphericall supficies was upon ye Axis of ye Eye.

Wren to Brouncker (30 July 1663), Royal Society MS EL.W.3 no.3.

Yet, we know from Wren's letter to Petty that this work was part of the programme for improving the telescope. Sprat also treats it as part of Wren's contribution to astronomy, and it is significant that Wren writes to Brouncker: 'ye wayes, by wch I did it, are too longe to rehearse; but ye projection in treble ye magnitude Sr. P. Neile may possibly find'. Wren went on to make an 'Artificial Eye, with the Humours truely & dioptrically made' (C. Wren, 1750, p. 198), to 'represent ye picture, as nature paints it' (Royal Society MS EL.W.3 no.3), which shows how completely he had adopted the attitudes of Kepler and Descartes.[16]

We know very little indeed of Wren's theory of light and refraction, but on the practical side of telescope improvement we have a little more information, and one of his proposed contributions to lens-grinding is particularly interesting.

Contemporary development of the telescope, as an instrument for qualitative observation, centred around the improvement of objectives of long focal lengths, so as to minimize the effects of spherical and chromatic aberration. These two types of distortion in the image stemmed, respectively, from the fact that a spherical surface – the only shape a lens could be given in practice – was not the theoretically correct form; and from the different characteristics of the coloured components of white light. Hooke once wrote, in the course of an attack on Cassini, that 'the Improvement of Telescopes, both for Length and Goodness . . . was first performed here by Sir Paul Neile, Sir Christopher Wren, and Dr. Goddard', adding that they employed and instructed the optical instrument-maker Richard Reeves and succeeded with good object glasses of 60-foot and 70-foot focal lengths, 'before any Mention was made of such being made in France' (Derham, 1726, p. 390). Comparison with what Hooke says elsewhere (Derham, 1726, p. 260) shows that here his patriotism has dulled his better judgment, but we do know that Wren and Neile had mounted what was, by all accounts, a good 35-foot telescope in 1658[17] and, about the same time, Neile produced a more doubtful objective with a 50-foot focal length.[18] Wren wrote a tract entitled 'A Method to make Telescopes with little Trouble and Expence, of great Length, to be used for any Altitutde' (C. Wren, 1750, p. 240) and he had considered, in certain circumstances, the possibility of dispensing with the tube (see Huygens, 1888, vol. 4, pp. 444–5; note also p. 433).

As well as increasing the lengths of telescopes, diaphragms that masked light from the edges of the lenses were important in reducing aberration, and were of interest to both Neile and Wren (for Neile note G. H. Turnbull, 1952, p. 120). Wren made use of the iris diaphragm (see Sprat, 1667, p. 214), though it has often been attributed to Hooke. It had, in fact, been suggested by Descartes (see Descartes, 1965, p. 122).

To attack the problem of spherical aberration at its source would involve a method for grinding aspherical lenses. Both Kepler and Descartes had recommended using lenses with hyperbolical surfaces, and the advantages of elliptical, parabolic and hyperbolic forms were generally accepted. The problem, of course, was to find a way of grinding and polishing glass to conform to any of these shapes.

Neile, Wren and Goddard were each directly involved in attempts to improve practical grinding techniques. Neile had tried unsuccessfully to make elliptical lenses (see Derham, 1726, p. 260), and Sprat wrote in 1667 that Wren had 'attempted, and not without success, the making of Glasses of other forms than Spherical' (Sprat, 1667, p. 314). There is other scattered evidence that Wren had been interested in grinding lenses prior to

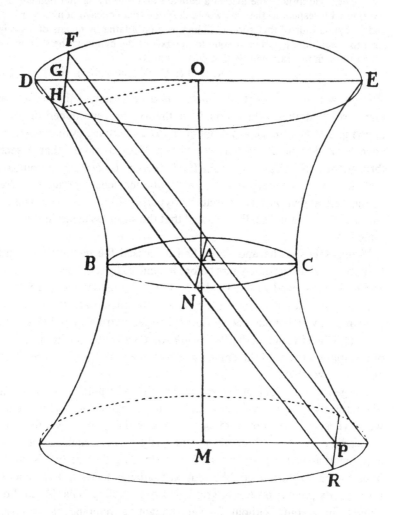

Fig. 2. The hyperboloid of revolution, (From *Philosophical Transactions*, 21 June 1669, vol. 4, no. 48, facing p. 961.)

this date,[19] and he was later to publish an ingenious idea for grinding hyperbolical lenses, the ideal form prescribed by Kepler and Descartes.

According to Wallis, Wren told him that the basis of the method first occurred to him

> as he had once seen a round wicker basket for sale in a shop, among other things, that was woven only from straight pieces of osier lying at oblique angles (and, I believe, crossing each other), the lateral surface of which displayed a cylinder hollowed from without [see Figure 2] (of that shape which with us is usually given to salt-cellars, or the sheaves of pulleys), he noticed that if one of those osiers led around the axis of a cylinder and preserving that oblique position with respect to the axis would describe that concavo-convex surface, and so cylindroids of that sort could be made on a lathe by means of a straight steel tool held in a position oblique to the axis of the cylinder, the section of which through the axis will be that curved line.
>
> A. R. Hall & M. B. Hall (1965), vol. 6, pp. 237–9

Wren saw that a straight cane acted as a generator of the surface and, therefore, that the surface could, in theory, be reproduced simply by applying a straight-edged tool obliquely to a revolving cylinder. He must have been delighted with his subsequent demonstration that a section through the axis was an hyperbola. If this account is correct,[20] it would seem that the search for a method of grinding aspherical lenses stimulated Wren's mathematical demonstration that the hyperboloid of revolution is a ruled surface, though it is usually assumed that the stimulus acted in the reverse direction.

Wallis said that he and Wren had discussed the method for grinding hyperbolical lenses some years before it came to light in 1669.[21] On 3 June 1669, Wren showed a model of his grinding engine to the Royal Society,[22] and the following week he explained both his engine and the geometrical principle on which it had been based.[23] The geometrical demonstration was first published in June in the *Philosophical Transactions* with only a 'hint' of the application, and the engine was described, though in no real detail, only in November.[24]

The geometrical demonstration, as a positive contribution to mathematics, has naturally attracted more attention than the engine, though the engine, as we have seen, was far from being an afterthought for Wren. This is misleading, since the breadth and integrity of his insight – theoretical, experimental and practical – is the outstanding feature of his work. A hyperbolical cylindroid could be cut on a lathe, but such a shape could not preserve its precise profile in grinding a hyperbolical lens. Wren did not simply throw out, without further thought, a reference to a possible application which was obviously of little real use. When Huygens objected that the 'hint' of an application to grinding lenses was of no value, Oldenburg realized that Wren would have to explain his model more fully in a second paper.[25] Wren himself replied that his machine was designed so as to allow for mutual wear, and that 'but for that he would have asserted nothing as to this business' (A. R. Hall & M. B. Hall, 1965, vol. 6, pp. 222, 306).

In Wren's proposed grinding apparatus (see Figure 3), two revolving cylinders, with their axes inclined, work against each other, to form two hyperboloids in contact along their common generator. At the same time, one of the cylinders also works against a revolving piece of glass, whose axis is set at right-angles to the two revolving cylinders. Based on Wren's geometrical demonstration, this application involves a more considered and sophisticated technique (for some contemporary reactions see A. R. Hall & M. B. Hall, 1965, vol. 6, pp. 396, 425–6, 448). Wren explained to the Royal Society that, by arranging three interacting bodies in this way, he hoped to overcome the problem of unequal wear,[26] and told Huygens: 'the glass and the machine will perfect each other' (A. R. Hall & M. B. Hall, 1965, vol. 6, p. 306). There is little record of the actual construction or materials used in the model or the proposed machine, since, typically, Wren was not prepared to spend time formulating a detailed description: 'operosa pictura & prolixa explicatione describere, mihi & artifici magnis fuerit molestum' (*Philosophical Transactions*, 15 November 1669, vol. 4, no. 53, p. 1060).

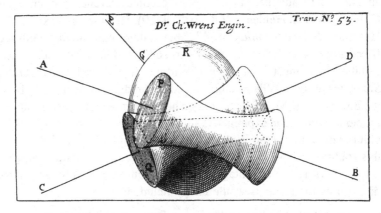

Fig. 3. Wren's engine for grinding hyperbolical lenses. (From *Philosophical Transactions*, 15 November 1669, vol. 4, no. 53, facing p. 1059.)

Wren had at least found a technique that was applicable in theory. Whether it was actually practicable was immediately questioned.[27] Hooke was asked to try the machine in practice and, after being repeatedly reminded, eventually decided that his own method was better.[28] Wren planned some trials himself, but probably never carried them out (see A. R. Hall & M. B. Hall, 1965, vol. 6, pp. 127, 448, 460).

Although unsuccessful, this is a very revealing example of the intimate relation between Wren's mathematical and practical insights, seen both in the genesis of his demonstration and in the subtle attempt at an application. As I have said before, our study of Wren's astronomy concerns the relation between astronomical advance and instrumental improvement; and if we pick up another thread beginning in theoretical astronomy, it leads us to a development in the telescope by which the latter could be used as an astronomical intrument. For the telescope was being developed, not only for qualitative observation but, also, for measurement.

When Wren spoke of his programme of lunar observations it was the quantitative aspect that he often stressed. The programme, we can date fairly accurately since, in September 1655, Wren and Wilkins were erecting a lunar telescope and, in the same month, Wren told Hartlib that he was working on a survey of the Moon's surface and a theory of her libration (see G. H. Turnbull, 1952, pp. 116, 114). On the problem of the Moon's libration, there is evidence that Wren made some progress but there is no record of his work.[29] His lunar survey produced more definite results.

Wren regarded his lunar survey, as he did his observations of Saturn, as an attempt to advance an area of telescopic astronomy opened up by Galileo (see Van Helden, 1968, pp. 219–20). His immediate stimulus was Hevelius's *Selenographia* of 1647,[30] but Wilkins had already published his *Discovery of a world in the Moone* in 1638, and his influence, indeed his collaboration, must have been important. In the tradition of Galileo's *Starry messenger*, Wilkins emphasized the fundamental similarity between the Earth and the Moon; he described her 'high mountains, deep vallies, and spacious plains', and explained Galileo's method for calculating the heights of lunar mountains (see Wilkins, 1802, vol. 1, pp. 63–74). Wren followed this tradition when, in 1657, he stressed that the telescope had granted astronomers a great advantage over the ancients and, among the telescopic discoveries the ancients would envy, included:

the Moon herself, that they should have a Prospect, as it they were hard by, discovering the Heighths and Shape of the Mountains, and Depths of round and uniform Vallies, the Shadows of the Mountains, the Figure of the Shores, describing Pictures of her, with more Accurateness, than we can our own Globe.[31]

Wren saw his contribution in the use of accurate measurement and, here, he was applying observational techniques which, though as yet unpublished, had already been developed in England. It is an interesting example of Wren drawing on the working astronomical tradition with which he was intimately involved.

In 1655 Wren said that his new selenography would be 'far more accurate than that of Hevelius, doing all by rule and demonstration which hath been hitherto done by guesses' (G. H. Turnbull, 1952, p. 114). He made the same point in 1656 when he wrote that the group at Oxford 'not only draw Pictures of the Moon, as Hevelius had done, but Survey her & give exact maps of her', and gave a clue to the source of this accuracy by saying that 'we make the Tube an Astronomical Instrūent to observe to Seconds'.[32] Previously, the telescope had been useful only for qualitative observations; now it would become a true astronomical instrument in the traditional sense – a device for making astronomical measurements.

For an accurate lunar survey Wren would have needed a method for measuring very small angular distances, both for plotting the Moon's features in plan, and for calculating the heights of lunar mountains. The latter involved measuring distances between points of light in the Moon's darkened portion and the boundary of the shadow. The problem of measuring small diameters came up at a Royal Society meeting on 9 January 1666/7, when Oldenburg read a letter from Adrien Auzout, 'mentioning a new method esteemed by him better than any hitherto practised, of taking the diameters of the planets to seconds'.[33] Hooke and Wren immediately pointed out that they had known of such methods long before, and were asked to set them down, 'that so it might be signified to the Parisian philosophers, that it was a thing not at all new among the English'.[34]

The key, of course, lay in using an eyepiece micrometer mounted in the focal plane of the objective. Auzout had a moving-wire micrometer, and it seems that Hooke used a similar technique (see King, 1955, pp. 98–9). Hooke also described using a graduated ruler (see Waller, 1705, p. 497), and he attributed the perfection of this method to Wren (see Derham, 1726, p. 272). It seems likely also that Wren used adjustable knife-edges or wires, since Sprat says that he 'added many sorts of Retes, Screws, and other devices to Telescopes, for taking small distances and apparent diamets to Seconds' (Sprat, 1667, p. 314; note also p. 250).

The principle was indeed 'not at all new among the English'. In about 1640 William Gascoigne had first applied cross-hairs to the focal plane of a telescope. He soon advanced to an eyepiece micrometer, using parallel knife-edges moved on a screw, and taking pointer-readings off a graduated disc (see King, 1955, pp. 94–7). The technique was known to associates of Gascoigne, such as Horrox, Crabtree and Oughtred,[35] and Ward may have been the link with the Oxford group. We saw that in the early 1650s he was 'procureing & fitting Telescopes and other instruments' for his observatory at Wadham. The instrument-makers whom he most likely employed were Christopher Brookes and Ralph Greatorex, both of whom were close to Oughtred.[36]

Wren's use of the micrometer in telescopes in the mid-1650s is an example of his drawing on a working tradition in which this development, though as yet unpublished, was known. It is sometimes said that Gascoigne's idea was lost, and the micrometer re-discovered later in the century, but this is not true. During the same period as his lunar survey Wren was also applying the micrometer to the microscope.[37]

He carried his work on the Moon a stage further by making a lunar globe. We first hear of it in May 1661, when the Royal Society conveyed to Wren a command from Charles II to make a globe of the Moon (see T. Birch, 1756, vol. 1, p. 21). But it seems that Wren had already been working on such a globe and that Neile and Moray had purposefully aroused the King's interest.[38] The Society were naturally pleased to have attracted Charles's attention, and anxious that Wren should satisfy him. They reminded Wren of his duty in July,[39] and the globe was completed in August.[40] It was made of pasteboard (see Monconys, 1665, vol. 2, p. 82), painted and carved in relief (see Sprat, 1667, p. 315) and, according to Moray, 'une tresiollie piece, artistement faite' (Huygens, 1888, vol. 3, p. 317).

Wren presented his globe directly to Charles and inscribed a dedication to him on the base.[41] According to *Parentalia:* 'His Majesty received it with particular Satisfaction, and ordered it to be placed among the Curiosities of his Cabinet' (C. Wren, 1750, p. 212). Here Huygens saw it in 1663 and found it 'fort plaisante a veoir avec toutes ses taches et petites vallees rondes' (Huygens, 1888, vol. 4, p. 369). It was also shown to both Sorbière and Monconys in 1663, and Sorbière confirms Charles's satisfaction when he says: 'His Majesty put me upon admiring it'.[42]

Along with the general praise for Wren's work, there was disappointment that his globe was rather small – only 10 inches in diameter – and Wren promised to make a larger, more accurate version for the Royal Society.[43] Three years later, in response to an inquiry from Hevelius, Wren was asked to 'defer no longer', and by then the Society were prepared to buy a globe for him to work on.[44] In January 1666/7 Wren was again reminded 'of the telescopical moon formerly promised by him' (T. Birch, 1756, vol. 2, p.

140). A plan was now devised for making a globe with the minimum effort on Wren's part. Moray would ask Charles for the loan of his globe, and Wren would employ a workman to copy the features onto a larger version: new observations could be added as they were made (see T. Birch, 1756, vol. 2, pp. 143, 154). The scheme began well: the Society instructed their own operator, and Neile reported that Wren had retrieved his earlier globe, but there is no record that another was ever made.[45]

The eyepiece micrometer had other, either real or proposed, applications in Wren's work, such as the quantitative study of the orbits of satellites of Jupiter and Saturn,[46] or an accurate map of the Pleiades.[47] It is relevant also to Wren's early project 'To find whether the Earth moves' (C. Wren, 1750, p. 198), which Sprat explains as a search for stellar parallax: 'He . . . propos'd Methods to determine the great doubt of the Earth's motion or rest, by the small Stars about the Pole to be seen in large Telescopes' (Sprat, 1667, p. 315).

Proposals, rather than attempts, were probably as far as Wren ever carried the idea, but we know that in the early 1650s he and Neile were discussing permanently mounting an objective at a considerable height, as part of a method for observing any variation in the meridianal altitude of a particular star. The technique may have been similar to Hooke's later application of a zenith telescope, but, as is so often the case, we can discover Wren's ideas only through the retrospective view of an associate. In 1663 Huygens wrote to Moray of his success in mounting a 35-foot telescope without a tube (see Huygens, 1888, vol. 4, p. 433). Moray at first replied that the method had not yet been tried in England, but recalled that, a long time since, Wren had had a similar notion for mounting a telescope of 60 or 80 feet. While Moray was writing his reply, Neile called on him and explained that, ten years previously, he and Wren had considered using a telescope without a tube, but that this had been only for a particular purpose:

ils n'ont pas songé à employer cette inuention pour l'usage ordinaire. seulement ils se proposoyent de placer un verre obiectif sur quelque grande hauteur et le fixer là pour obseruer et comparer les altitudes meridionales de quelque estoile afin de tascher de decouurir sil y a quelque paralaxe.

Huygens (1888), vol. 4, p. 444–5.

It was in 1669 that Hooke began his attempt, and mounted a micrometer with the eyepiece of a zenith telescope, which he set up, without a tube, in his rooms at Gresham College (see Hooke, 1674, *An attempt to prove the motion of the Earth from observations* (in Hooke, 1679, which is reprinted in Gunther, 1920–67, vol. 8)). On the basis of inadequate results, he announced the discovery of a sensible parallax 'and consequently a confirmation of the Copernican System against the Ptolomaick and Tichonick' (Hooke, 1674, in Hooke, 1679, p. 25). His contemporaries, however, were right to be sceptical,[48] and Hooke himself was not altogether happy – he would have preferred a taller and more stable mounting (see Hooke, 1679, pp. 22–3). About the same time, Hooke and Wren were working on the design of the Monument in London, which was 200 feet high and, according to James Hodgson, who was a relation of Wren's, built hollow in order to serve as a zenith telescope for discovering stellar parallax.[49]

The Monument was not sufficiently rigid for such accurate work but, in February 1703/4, Wren proposed to make use of the massive structure of St. Paul's (see Royal Society MS Council Minutes, 2 (copy), p. 169). Constantine Huygens had given the Royal Society the lenses of a 123-foot telescope in 1692 'with the Apparatus for using them without a Tube',[50] but it had proved difficult to use them effectively. Wren's plan was to mount the telescope in the south staircase at St. Paul's, and Hodgson, who on Wren's recommendation had been employed and trained by Flamsteed at the Royal Observatory (where zenith observations had also been attempted), was to have made the observations.[51] Again the plan failed: Huygens's telescope was, in fact, too long.[52]

The application of telescopic sights to existing techniques of precision astronomy was a natural complement to Wren's interest in the micrometer. John Ward tells us that Wren donated to the Savilian Library 'a large brass quadrant placed in a standing frame, with a radius of about 26 inches, and furnished with two telescopic sights'.[53] When Wren himself listed the instruments necessary for an observatory in 1681, he mentioned 'a large murall Quadrant fixt to a wall trewly built in the meridan', 'a pole to rayse large Telescopes & manage them' and 'A Quadrant to take distances fixt to a Foot soe as it may turne to all sort of plains'. Typically, this instrument should have 'Telescope sightes & many nice Joynts & Screwes' (Caröe, 1923, p. 31).

Whether the use of telescopic sights led, in practice, to more accurate observations, was disputed between Hevelius and the English school, represented by Hooke and Flamsteed. When Hevelius claimed that telescopic sights had never been tried on large instruments, Hooke replied that he had used several,

and particularly one of Sr. Christopher Wren's invention, furnished with two Perspective Sights of 6 foot long each, which I made use of for examining the motions of the Comet, in the year 1665.

Hooke (1679), p. 77.

This was Wren's 'double-telescope' which Hooke describes elsewhere in the *Animadversions* (1674), as

two square Wooden Tubes, joyn'd together at the end next the Object by a Joynt of Brass, and the Angle made by the opening of them, measured by a straight Rule.

Hooke (1679), p. 54; note also p. 32.

Both Sprat and Monconys also describe this instrument[54] and Hooke gives a very detailed account in a lecture that was probably written in the period 1665–9 (see Waller, 1705, pp. 498–503).

The instrument was used by two independent observers. When each had located his star or planet exactly on the reference point, the arms were locked and the angle read off a straight scale between them. It was light, easy to use, could be folded away and easily transported and was accurate, Hooke claimed, to minutes of arc: 'a worthy Product of its excellent Inventor'.[55]

However, it seems likely that the double-telescope is one of the links between Wren's practical astronomy and the search for a solution to the navigational problem of finding longitude at sea. We will therefore meet it again in the next chapter. Indeed, the importance of instrumentation has been one of the most striking features of the whole range of Wren's work in astronomy, and this will be emphasized further when we look specifically at the problem of longitude. This is, of course, just what we would expect from Wren's background and training in the established tradition of the mathematical sciences.

It is especially clear that Wren moved easily between theoretical and practical problems, and that these were complementary and not separate aspects of his talent. Their intimacy is best seen in the proposal for grinding hyperbolical lenses for telescopes, but a great deal of his astronomy was linked closely with the development or improvement of the telescope. The most lavish compliment that Hooke paid Wren in print has often been quoted in isolation from any context, but it is only after looking at the work and background of both men that we can appreciate its true significance:

since the time of Archimedes, there scarce ever met in one man, in so great a perfection, such a Mechanical Hand, and so Philosophical a Mind.

Hooke (1665), preface.

5

Longitude

The single, most important technological problem of the seventeenth century was to discover a practicable method of finding longitude at sea. The word 'technological' is used here with care. Other critical technical problems existed, of course, but a solution to the problem of longitude was deliberately sought through developments in science or natural philosophy. Both the educated navigator and the Baconian natural philosopher expected that a solution, when it came, would develop from an application of some scientific principle. In this sense, the problem was understood to be one of technology.

Most proposed solutions relied on some method for finding the time at a standard meridian, distant from the observer. This might be done by carrying a timepiece set to the standard time, or by observing the progress of some celestial phenomenon such as the motion of the Moon or of the satellites of Jupiter, and consulting tables constructed for time at the standard meridian. The longitude difference could then be found merely by comparing standard time with local time – readily discovered from Sun or stars – an hour's difference in time being equivalent to 15 degrees of longitude. Alternatively, improvements might be made in the traditional methods of keeping an account of the voyage in bearings and distances (so-called 'dead reckoning'), or an absolute measure of position on the Earth's surface might be found from some terrestrial phenomenon such as magnetic variation.

No working solution emerged during the century; the navigator in 1700 was as much in the dark as regards his longitude as his counterpart had been in 1600. But the problem itself had been a powerful stimulus in natural philosophy, and its story is linked with many of the major developments in the physical sciences. Galileo's telescopic observations of Jupiter's satellites immediately suggested a possible solution, and Roemer's discovery of the finite velocity of light was made in pursuit of this idea. Throughout the century, much of the work on terrestrial magnetism was motivated by the search, and Halley's research, in particular, seems to represent the real beginning of the whole subject of geophysics. The most important and far-reaching developments in horology, in the work of Hooke and Huygens, were linked to the goal of a viable marine timekeeper. The first national public observatories were established to provide the basis of a solution, and the attempt to wrest an accurate lunar theory from Newtonian dynamics had the same end. Wren said of the longitude in 1657, 'former Industry hath hardly left any Thing more glorious to be aim'd at in Art' (C. Wren, 1750, p. 206).

The areas that excited most attention, at this interface between the science and technology of the seventeenth century, were Jupiter's satellites, a mechanical timekeeper, terrestrial magnetism and the proposed method of lunar distances. (A useful explanation of these methods is in Gould, 1960, introduction.) Wren turned to each of them in his running battle with the problem. Indeed, the longitude is the most recurrent theme in his work, and spans almost the whole of his adult life. As we would expect in someone educated within the English tradition in mathematical sciences, it was part of a broader interest in navigation and nautical science, but Wren reflects the situation of his time by devoting most attention to the longitude.

The Royal Society were quick to realize the potential benefits of pursuing nautical science. On 19 December 1660 it was ordered:

The Dr. Petty and Mr. Wren be desired to consider the philosophy of shipping, and to bring in their thoughts about it to the society.

T. Birch (1756) vol. 1, p. 7.

In the years that followed, Wren seems to have been a natural choice for Royal Society committees dealing with maritime questions. The 'philosophy of shipping' would have included the design of vessels and of sails, in addition to navigation. Petty's interest at this time was in ship design[1] and the Society's President, Viscount Brouncker, was later asked specifically to consider this question.[2]

Naval architecture, as one of the branches of architecture (a classification which Wren himself accepted (see Bolton & Hendry, 1923, vol. 19, p. 140)), was a subject within the mathematical sciences. Wren's early proposals included 'New Ways of Sailing' and 'Fabrick for a Vessel for War' (C. Wren, 1750, p. 198) and, according to Sprat, he worked on 'laying down the Geometry of Sailing, Swimming, Rowing, Flying, and the Fabrick of Ships' (Sprat, 1667, p. 316). There is sufficient evidence to confirm Wren's active interest in naval architecture,[3] but no evidence at all that he made any significant contributions.

It seems that Wren had more success in dealing with the mechanical principles of sailing against the wind, although no actual account has survived. However, Sprat's remark that on the question 'to what Mechnical powrs the Sailing (against the wind especially) was reducible; he shew'd it to be a Wedge' (Sprat, 1667, p. 316), agrees with the practice Wren adopted, elsewhere, of epitomizing a theory in the properties of a simple machine. His theory of elastic impact was similarly presented in terms of a balance (see Chapter 7). It is further typical of Wren that he went on to construct 'an Instrument, that Mechanically produc'd the same effect, and shew'd the reason of Sayling to all Winds' (Sprat, 1667, p. 316). The earliest definite reference to this instrument shows that in 1663 it was already in use by Wren's earlier colleague on the nautical committee of the Royal Society, William Petty. At a Society meeting on 11 February, after a report of trials of a new ship designed by Petty, John Graunt said that 'in a letter to him from Sir William it was mentioned, that the ship went as near the wind, as by Dr. Wren's brass instrument it was possible' (T. Birch, 1756, vol. 1, p. 194). The instrument later found its way into the Society's museum.[4]

Wren's interest in the mechanics of rowing followed a similar pattern, the simple machine in this case being 'a Vectis or a moving or cedent Fulcrum', with instruments being designed to investigate resistance to motion in a liquid medium.[5]

While there are scattered references to other nautical ventures, such as a newly-designed sounding instrument,[6] most of Wren's maritime interests can be subsumed under the general problem of finding longitude at sea, whose relevance ranged very widely indeed. An entry in the *Parentalia* list of Wren's contributions to the work of the Oxford group sums up the relevant and inter-connected subjects as: 'The best Ways for reckoning Time, Way, Longitude and observing at Sea' (C. Wren, 1750, p. 198).

It seems that one of Wren's earliest approaches to the problem was through his experimental study of terrestrial magnetism, which dates from about 1656–7. Sprat relates this work specifically to navigation, and describes one particular experiment which involved tracing the direction of 'Magnetical virtue' in the plane containing the poles of a terella or spherical loadstone (see Sprat, 1667, pp. 315–16). The apparatus was given to the museum of the Royal Society, and Grew gives a more detailed description, mentioning not only steel filings as Sprat does, but also 32 needles for tracing 'the different respects of the Needle to the several Points of the Loadstone' (Grew, 1681, p. 364). Grew also includes this instrument under navigation.

Clearly, this work relates directly to the magnetical experiments of William Gilbert who had a great influence on natural philosophy in

England in the first half of the seventeenth century (see Chapter 6). Wren's inaugural lecture (1657) at Gresham contains a powerful eulogy on Gilbert and his work, the chief immediate result of which was 'an admirable Correspondence betwen his Terella, and the great Magnet of the Earth' (C. Wren, 1750, p. 204). Wren was fully aware too of the later work on terrestrial magnetism, carried out by the Gresham Professors Gunter and Gellibrand, which resulted in the discovery of the secular changes in magnetic variation. Thus, in the same speech, he says that:

the whole Doctrine of Magneticks, as it was of English Birth, so by the Professors of this Place was augmented by the first Invention and Observation of the Mutation of the magnetical Variation.

C. Wren (1750), p. 206.

Wren's terella apparatus was probably devised in connection with his experiments for measuring, directly, what Gellibrand had called 'that abstruse and admirable variation of the Variation' (Gellibrand, 1635). In about 1656 Wren wrote to Petty: 'I have lately caused a Needle to be made of 40 Inches, by which I hope to discover the Annual Motion of Variation & Anomalies in it' (Bennett, 1973, p. 147). And, in 1657, Boyle told Hartlib that he was experimenting with Wren on the variation (see G. H. Turnbull, 1952, p. 112). Wren wrote some sort of treatise entitled: 'To observe the Variations of the magnetical Needle.'[7]

The discovery of secular changes in variation might have been expected to repudiate earlier schemes for discovering longitude from variation measurements. But this was not the case. Halley's extensive work, later in the century, was aimed at a longitude method. It was thought that alterations in variation could be reduced to some kind of theory, and this explains Wren's hope that he might 'discover the Annual Motion of Variation & Anomalies in it'. He said in 1657 that the science was

a Thing, I confess, as yet crude, yet what may prove of Consequence in Philosophy, and of so great Use, possibly to the Navigator, that thereby we may attain the Knowledge of Longitudes

C. Wren (1750), p. 206.

He was later concerned that an azimuth compass, adapted for convenient use at sea, should be devised (see T. Birch, 1756, vol. 4, p. 102).

The consequences, in philosophy, are more obscure but may refer to Wren's contemporary interest in the role of 'Magneticks' in explaining planetary motion (dealt with in Chapter 6) since, at the same time, he described the subject as 'a Kind of Terrestrial Astronomy, an Art that tells us the Motions of our own Star we dwell on'.[8]

Wren was shortly at work on a formal treatise on finding the longitude, and had switched his faith to one of the other general areas where solutions were sought. An unpublished reference in the earliest manuscript of *Parentalia* is to a tract, written at Oxford, and dated by Wren's son to about 1658. He even quotes the opening of Wren's preface, but breaks off before reaching anything of substance:

A Discourse in the style of a Letter on that grand Quaesitum in Navigation, the finding of Longitude at Sea frequently and certainly. – 'After the Labours of so many Witts (says the Preface) as good as frustrate, 'twill seem a great Presumption to think it fesable: our greatest incouragement is that the World hath yet gone the wrong way to work. 'Tis not from Astronomy nor Magneticismes, but from more obvious Principles we must receive this Magnale Artis. tc^{ra}

British Library MS Add. 25 071, fo. 89.

From Wren's subsequent interests, it seems likely that the method based on 'more obvious principles' involved either a log for measuring (and perhaps recording) a ship's velocity or distance travelled, to render the traditional 'dead reckoning' more reliable; or, alternatively, a marine clock for carrying standard time.

Scattered references to Wren's log or 'way-wiser' for use at sea are all that survive. A letter read to the Royal Society in March 1684/5 prompted Hooke to observe that, 'As to the instrument for finding the way of a vessel, he could not see, that it differed from Sir Christopher Wren's' (T. Birch, 1756, vol. 4, p. 378), and we know that Wren wrote a tract, 'To find the Velocity of a Ship in sailing' (C. Wren, 1750, p. 240). As is so often the case, Hooke's parallel interests are enlightening. He too devised a marine way-wiser in 1683, and pointed out that it could also be used to measure the velocities of a flowing river at different points and, in turn, 'the quantity of the water vented by any river into the sea' (T. Birch, 1756, vol. 4, pp. 230–1). And Wren had earlier reported (in 1670) that in surveying two rivers 'there had an estimate been made by him of the quantity of the water, which they hold, by the velocity and solidity of the rivers'.[9]

In July 1714 the British Government offered a considerable reward for a practical method for finding longitude at sea within certain specified limits. It was, no doubt, in response to this that in November of the same year Wren tried to stake a claim, in a manner familiar to a seventeenth-century scientist, for the methods he was pursuing. He presented a sealed paper to the Royal Society, together with a letter to Newton as President:

I Present to the Royal-Society, a Description of three distinct Instruments, proper (as I conceive) for Discovering the Longitude at Sea: They are describ'd in Cypher, and I desire you would, for Ascertaining the Inventions to the

Rightfull Author, Preserve them among the Memorials of the Society, which in
due time shall be fully explain'd by . . . Chr: Wren.

British Library MS Add. 25071, fo. 115.

Brewster found a copy, by Halley, of Wren's cyphers among Newton's
papers, and they were decoded.[10] One ('Pipe screwe moving wheels from
beake') relates to a log, fixed to the prow or 'beake' of the ship, with the flow
of water turning a helical pipe. It is interesting that Wren was still at work on
such a method, since Hooke had realized the difficulty of distinguishing
between the ship's motion and currents in the ocean (see Waller, 1705, p.
561).

A second cypher ('Wach magnetic balance wound in vacuo') relates to
the most obvious method of all, that of carrying standard time by a marine
chronometer, the difference between the time at a standard meridian and the
local time being a direct measure of difference in longitude. Much of the
practical development in horology during the seventeenth century was
aimed at a reliable sea-clock. It is difficult to guess at Wren's thinking in this
instance, though the notion of placing a watch in a vacuum goes back at least
to a Royal Society meeting of 12 February 1661/2, when 'Dr. Wren
proposed to try a watch in Mr. Boyle's engine' (T. Birch, 1756, vol. 1, p.
76). Boyle's own similar experiment related to the question of whether
sound can be propagated in a vacuum (see T. Birch, 1756, vol. 2, p. 500),
but the context of Wren's suggestion shows that he was concerned with the
performance of the watch (compare T. Birch, 1756, vol. 4, p. 183).

Wren's interest in mechanical horology had a long history. According to
Hooke:

Dr. Wren, Mr. Rook, Mr. Ball & others made use of an Invention of Dr.
Wren's for numbering the vibrations of a pendulum a good while before Monⁿ
Zulichem [Huygens] publisht his.

Robertson (1931), p. 168.

This would date the device to Wren's earlier period at Oxford but, quite
apart from the problem of Hooke's enmity for Huygens, an instrument for
numbering vibrations is not necessarily a pendulum clock.

Sprat gives some results of Wren's study of pendulum motion, such as
that the velocity of a simple pendulum bob is related to time by a sine
function.[11] Wren was interested in vibrative motion in various forms – the
undulation of mercury in a curved tube (see T. Birch, 1756, vol. 1, p. 115);
the vibrations of a glass bell (see T. Birch, 1756, vol. 4, pp. 46, 48); and the
motion of water waves as revealed by the bobbing of a cork (see T. Birch,
1756, vol. 4, p. 71).

Hooke discussed with Wren, not only his work on pendulum clocks, but also that on spring-regulated watches.[12] On 27 February 1674/5, for example, he 'Discoursed much with Sir Ch. Wren about spring watch', and on 21 March 1675/6 they 'discoursed about watches for pocket and for Longitude' (H. W. Robinson & W. Adams, 1935, pp. 150, 221). One purpose for Wren's constant-temperature furnaces (an idea that interested Hooke also) was 'keeping the motion of Watches equal, in order to Longitudes and Astronomical uses' (Sprat, 1667, p. 317). Another idea of Wren's that Hooke developed, was that a lamp designed to supply a constant flow of oil could be adapted as a form of clepsydra.[13]

A further aspect of Wren's interest in the pendulum concerns the idea that the length of a seconds pendulum could form the basis of a universal standard of length. Sprat, Wilkins and Hooke all attribute the first suggestion to Wren, and Hooke dates it to some years before the foundation of the Royal Society.[14] To prepare 'the pendulum experiment' was one of the earliest instructions the Society issued to Wren (see T. Birch, 1756, vol. 1, pp. 4, 7), and he was a member of later committees set up to consider the related ideas of Brouncker and Huygens.[15] In spite of doubts about such imponderables as the constancy of gravity, Wren still referred in February 1677/8 to the length of a seconds pendulum as 'the universal standard . . . which was the same all over the world, and would ever be so in all ages' (T. Birch, 1756, vol. 3, p. 384).

An alternative source of a universal measure that interested Wren was the length of a degree of latitude on the Earth's surface.[16] This seems to have been the idea he favoured in 1682, though he raised the objection of the possible aspherical shape of the Earth (see T. Birch, 1756, vol. 4, pp. 150–1). One of his later tracts was entitled: 'Of the true Shape of the Superficies of the terrestrial Globe'.[17]

It is natural that astronomical methods should figure largely in Wren's attack on the longitude problem. A recurrent or progressive and measurable celestial property, reduced to a theory, could provide the navigator with a standard clock, against which to compare the measurable local time. The lunar motion through the fixed stars, was, of course, one of the more obvious candidates. Noting occultations of stars by the Moon, or measuring the distance of the Moon from a star, could provide the raw data for determining time at a standard meridian. Unfortunately, the ingredients necessary to the method – an appropriate instrument for taking angular measurements at sea, an accurate lunar theory and an accurate chart of stars in the Zodiac – did not exist. Wren was a member of the commission set up in 1672 by Charles II to consider a proposed lunar method for longitude (see Baily,

1966, pp. 37–9) and the deficiencies that were then pointed out, so impressed the King that the Royal Observatory was established in 1675 specifically to remedy them.

Wren's own interest in some form of lunar method went back at least to some ten years prior to the commission. A fragmentary manuscript note[18] reveals some relevant dialogue between Wren and the Earl of Sandwich in 1662, but little of substance. In April 1663, however, Moray (later a prime mover in the establishment of the Royal Observatory) proposed to the Royal Society a project for fixing the positions of the stars in the Zodiac (see T. Birch, 1756, vol. 1, p. 219). Brouncker, Moray, Ball, Wren, Pope, Croone and Hooke were asked to choose areas for their parts in the survey, and Hooke and Wren decided to collaborate in surveying Taurus.[19] In June, Monconys saw Wren's 'double-telescope' at a meeting of the Society, describing it as 'vu instrument qu'a fait M. Renes pour prendre la distance de deux estoiles' (Monconys, 1665, vol. 2, p. 73). This is the instrument described in Chapter 4, as an early example of the application of telescopic sights to measuring instruments. Monconys himself, particularly noticed the reference points that allowed the exact centres of the stars to be fixed. He also tells us that one tube was mounted on a wooden stand in such a way that it could swivel, as the other was moved, without changing its direction.

In the context of its time it seems very likely that this whole project, including the instrument provided by Wren, should be seen as a fresh attempt on the problem of the longitude. Hooke, Wren's associate in the proposed survey, places the double-telescope very firmly in this context. In the lectures, where he described Wren's instrument in detail, Hooke proposed a vast observational programme for establishing the positions of the fixed stars and providing a basis for true theories of the motions of the Earth, Moon and planets. This would, of course, take advantage of the new technology and consequent improvement in accuracy, available through the use of telescopic sights. The role of the double-telescope was 'for making a compleat Hoop or Zone of all the fix'd Stars in the Zodiac' (Waller, 1705, p. 498), which, in conjunction with the true lunar motion, would be the astronomical basis of a method of finding longitude at sea (see Waller, 1705, p. 500). Wren's own written account of his instrument embraced a similar scope. Its title has survived as: 'Description of an Instrument for the observing Distances of fix'd Stars, and the Planets, and Appulses to the Moon; by two Telescopes join'd like a Sector, so as to give the true Angle of their Distances' (C. Wren, 1750, p. 240).

It was also in the early 1660s that Wren devised a graphical method of calculating the time and extent of a solar eclipse, a method which could be applied to occultations of fixed stars by the Moon (see Flamsteed, 1680, preface). Since no calculation was needed, this would have been particularly appropriate for use at sea, as Flamsteed recognized when he arrived at the same construction in 1676. When Flamsteed's work was presented at a Royal Society meeting, Wren revealed his own and produced 'a like Projection neatly drawn on Pastboard, and fitted with several Ingenious contrivances of Numbers and Scales for the Construction of Solar Eclipses in our Latitude' (Flamsteed, 1680, preface). Wren's idea had come to light in a typically retrospective manner, and it was, no doubt, in connection with this revived interest that on 16 September 1676, Hooke recorded that Wren 'told me his way of calculating Eclipses of the Sun by an instrument' (H. W. Robinson & W. Adams, 1935, p. 250).

The second astronomical method of finding longitude at sea, which was propounded from time to time in the seventeenth century, was based on observing the eclipses of Jupiter's satellites. (Wren himself referred to their 'Multitudes of Eclipses' in 1657 (C. Wren, 1750, p. 205).) The idea, which had occurred already to Galileo, was that if tables of predicted times of eclipses were established for a standard meridan, Jupiter and her satellites could become a standard clock for mariners throughout the world. In about 1656 Wren said that the eyepiece micrometer was enabling the astronomers at Oxford to make an accurate study of the motions of these satellites (see Bennett, 1973, p. 147) and this work was continued at Gresham by Wren's mathematical colleague there, Laurence Rooke.

Rooke's extensive study and many observations were aimed at a theory of the satellites' motions and, in turn, at a method of finding longitude on land.[20] Walter Pope tells how he himself sought help at the Royal Society when Rooke became too ill to complete his work;[21] and after he died on 27 June 1662, Pope and Wren were asked, on 9 July 'to continue the observations of Jupiter's satellites' (T. Birch, 1756, vol. 1, p. 88). Wren was reminded 'of prosecuting Mr. Rooke's observations' in September, and Pope was asked to remind him again in February 1662/3 (see T. Birch, 1756, vol. 1, pp. 109, 194). From the contemporary accounts of both Sorbière and Huygens, we know that the purpose was a method for finding longitude and, as Huygens specifically says, a method for use on land.[22] But it seems that little was done, although Ball, Hooke and Wallis were all asked to help (see T. Birch, 1756, vol. 1, pp. 194, 216, 220, 328). Pope says that, at his death, Rooke's papers passed to Seth Ward, and in April 1663 Pope was asked by the Royal Society to retrieve them.[23] A similar request was made directly to Ward in October 1666 (see T. Birch, 1756, vol. 2, p. 116), but when Pope came to write his *Life of Seth Ward* in 1697, he said that the papers 'for ought I know, have since perisht' (Pope, 1697, p. 117).

The difficulty with using Jupiter's satellites to find longitude at sea was that a fairly long telescope, unmanageable on board ship, was needed to define an eclipse with any precision.[24] However, we know that Wren later tried to overcome this by devising new observational techniques. In December 1704 David Gregory records:

Sir Christopher Wren says that he doubts not but a 5 foot Telescope made (by two reflecting plain mirrors) twenty inches long and braced to the head & orbite of the Eye, might at sea discover Jupiters Satellites.

Hiscock (1937), p. 23.

Wren based his claim on some rough trials in a moving coach, and he was applying the method of shortening telescopes, which had been suggested by his friend Hooke. Coupled to the telescope proper was a finder, viewed by the other eye, so that Jupiter could be located quickly. This, or something like it, was the third instrument of 1714, for the final decoded cypher reads: 'Fix head hippes handes poise tube on eye'.

Each cypher relates to a method that had interested Wren for a good many years, and it seems likely that in 1714 he was concentrating on the instrumentation needed to apply them effectively. The cyphers were intended to secure Wren's claim to the instruments themselves. At the same time he wrote a long treatise, 'with various Problems, Demonstrations & explanatory Figures engraven on Copper Plates, still improveing' (British Library MS Add. 25 071, fo. 115). Still improving – Wren was not satisfied even then, and when his son began work on *Parentalia,* Wren was still at work on the longitude. Sometime after 1719, his son wrote:

the Subject has been reassum'd by him of late years, and it is to be hoped may, for the benefit of Mankind, be brought to Perfection, before his death, if not done allready.

British Library MS Add. 25 071, fo. 89.

When John Ward, biographer of the Gresham professors, examined Wren's longitude papers, he concluded that: 'Sir Christopher had his thoughts very early upon that subject, and always kept it in his view afterwards.' Some he dated to 1660, others to 1720, 'and the whole consists of divers methods proposed by Sir Christopher for that end, with draughts of several instruments proper for the purpose'.[25]

Wren's son may even have hoped that a solution could still be found among his father's papers, for it was probably in this connection that he wrote to James Hodgson in 1739: 'Pray what do you make of the Longitude, I should be curious to know (British Library MS Add. 6209, fo. 205). It had indeed been the single most sustained problem that had concerned his father. In practical terms, Wren was no more successful than any of his contemporaries; but for him, along with others, the longitude had been a stimulus, if not to improvements in navigational technique, at least to significant developments in natural philosophy.

6

Cosmology

One of the most enduring intellectual landmarks of the seventeenth century was Newton's *Philosophiae Naturalis Principia Mathematica* of 1687. Historical interest in the period's cosmology has naturally been dominated by the *Principia* – the evolution of Newtonian dynamics, its application in Newton's 'System of the World', and its troubled reception in contemporary natural philosophy.

Glimpses of the book's prehistory are found in the text itself. In introducing the mathematical concept of a centripetal force, varying inversely as the square of distance from the attracting centre, Newton mentions that such a force applies in reality in the case of heavenly motions, 'as Sir *Christopher Wren*, Dr. *Hooke*, and Dr. *Halley* have severally observed' (Newton, 1947, p. 46). This little remark was the sole concrete result of much passion and ill-feeling, and its genesis is perhaps a useful introduction to that aspect of cosmological theory in the thirty or so years before the *Principia* in which Wren had a hand.

When, in May 1686, Robert Hooke learnt the basic principles of the cosmology that Newton would announce in his forthcoming book, he became anxious that his part in its development should be properly acknowledged. In a series of letters, written between November 1679 and January 1679/80, he had indeed presented Newton with the basic conceptual ingredients of the Newtonian cosmological programme: that a planetary orbit is the resultant effect of a rectilinear inertial motion and a centripetal attractive force governed by an inverse-square distance law. Having failed to find a demonstration himself, he had appealed to the mathematical prowess of the Lucasian Professor at Cambridge.

How much Newton owed to Hooke's correspondence need not detain us here.[1] Of this, Hooke himself had no doubt, and he was particularly concerned that the inverse-square relation should be acknowledged as due to him. The rest of the programme he had already announced in print anyway. It was Halley who had set himself the delicate task of seeing

Newton's manuscript through the press, and who had to report Hooke's exceptions. Newton parried by introducing Wren. He told Halley that in about 1677 he had visited Wren and 'discoursd of this Problem of Determining the Heavenly motions on philosophicall principles' (H. W. Turnbull *et al*, 1959, vol. 2, p. 433). He seemed to remember that Wren was then already acquainted with the inverse-square relation, and he asked Halley to check this with Wren himself. The point was that then Hooke could not claim even second place behind Newton (see H. W. Turnbull *et al*, 1959, vol. 2, p. 435).

Halley's reply (see H. W. Turnbull *et al*, 1959, vol. 2, pp. 441–2) related something of his own experience – that Wren had, in 1684, encouraged both Hooke and himself to produce a demonstration of the planetary motions from these principles, but without success on either side. Also, he had called on Wren and he reported Wren's account of his own ideas. However, to appreciate more fully the significance of Wren's reply, we must look first at the conceptual background to the problem in the seventeenth century.

The problem of the efficient cause of planetary motion had been precipitated, if not specifically posed, by Tycho Brahe's denial, on empirical grounds, of the solid orbs of the Aristotelian cosmology; and the search of an alternative physical account had a long history in English astronomy. In Book 6 of his *De magnete* of 1600, William Gilbert, having already demonstrated that the Earth has the properties of a large spherical magnet, moved on to deal with the wider role of magnetic influences on a cosmic scale.[2] Without specifically committing himself to the Earth's annual orbit, postulated by Copernicus, he gave a 'proof' of the diurnal rotation by analogy with the property attributed to a spherical loadstone, of rotating spontaneously when its axis is inclined to the plane of another magnetic influence. For Gilbert, the source of this motive magnetic influence was the Sun; and the Earth is, of course, inclined to the ecliptic – the plane containing the Sun's apparent orbit. By extension, Gilbert envisaged the magnetic influence of the Sun controlling the motions of the other planets and, thereby, introduced a key concept of cosmic, magnetic action-at-a-distance, a concept that would be crucial to later developments.

Gilbert understood his magnetic influences as spiritual in nature; they were thoroughly animistic, the 'magnetic' properties of the Earth being the expression of an Earth-soul. Yet, equally, his method was thoroughly experimental, and the power of his cosmology was his identification of cosmic influences with the magnetic forces which he could experiment with on Earth. The latter, incidentally, were also the evidence of ensouled material. Because of his thoroughgoing experimentalism, it was relatively easy for Gilbert's cosmic influences to shed the animistic interpretation he had given them, and thus to become physical forces – a straightforward, indeed characteristic, mode of material action.

We can see this transition in Kepler's use of Gilbert's ideas. It was Kepler who reinterpreted the role of astronomy by insisting that the astronomer must not simply devise a geometrical model that would reproduce the observed appearances, but must also give a physical account that would explain them. In his *Mysterium cosmographicum* of 1596, the planets were moved by a soul located in the Sun, but in the *Astronomia nova* of 1609, the motive influence whose source was again the Sun, was a physical, magnetical force. This key concept had come from the *De magnete*.

In England also, Gilbert had a powerful influence on the theoretical development of Copernicanism. His 'proof' of the Earth's diurnal rotation helped to establish the new astronomy, and it is significant that the common conservative position in the cosmological debate in England was the so-called 'semi-Tychonic' system, where the Earth had a diurnal rotation, but where the Sun, with the planets, made an annual revolution about the Earth (see Johnson, 1937, chapter 7). It is further significant that it was in England that Kepler's work excited most interest.[3]

A number of English astronomers tried to develop physical explanations of planetary motion, by applying Gilbert's notion of cosmic magnetical force. F. R. Johnson has pointed to 'a prevailing eagerness to evolve some new physical explanation of the movements of the planets in their orbits' (Johnson, 1937, p. 235), and has shown how Gilbert had provided a general conceptual foundation for such speculation. This theoretical tradition was maintained through the informal network of the English scientific community, and had a direct influence on Wren. This was an important survival for, by the mid-century, the Cartesian mechanical philosophy was fast becoming the dominant natural philosophy in continental Europe, and was making important inroads in England. Cartesian philosophy severely restricted the theoretical vocabulary of cosmology; for all material action, to be intelligible at all, had to be reduced to action on contact – direct collision or pressure transmitted through a medium. Other forms of material action were proscribed on fundamental metaphysical grounds, so that their survival in the English debate – in one form or other – became especially important.

If we take the example of Wren's philosophical mentor, John Wilkins, Gilbert's influence is clearly seen in Wilkins's two popular astronomical works, the *Discovery of a world in the moone* (especially in the edition of 1640) and the *Discourse concerning a new planet*. Gravity is discussed in the *Discovery* in terms of a magnetic attractive 'vigour', distributed, as Gilbert had argued in the case of a spherical lodestone, in a spherical field around the Earth. The analogy between gravitational and magnetic attractions is used freely, though, unlike Gilbert, the distinction between them is now carefully drawn.

This great globe of earth and water, hath been proved by many observations, to participate of Magneticall properties. And as the Lodestone dos cast forth its owne vigor round about its body, in a magneticall compasse: So likewise dos our earth. The difference is, that it is another kind of affection which causes the union betwixt the Iron and the Loadstone, from that which makes bodies move unto the earth.

Wilkins (1640), pp. 213–14.

Further, Wilkins does have the idea of a distance-related law. Though he uses the term 'magneticall' in the loose sense, characteristic of English cosmological discussion, Wilkins is writing of gravitational attraction when he says:

you must not conceive, as if the orbe of magneticall vigor, were bounded in an exact superficies, or as if it did equally hold out to such a determinate line, and no farther. But . . . it is probable, that this magneticall vigor dos remit of its degrees proportionally to its distance from the earth, which is the cause of it.

Wilkins (1640), p. 232.

Ideas derived from Gilbert's cosmic magnetism appear also in the *Discourse,* first published in 1640 (see Bennett, 1981).

The influence of the Cartesian philosophy was just beginning to be felt in England. By 1654 Wilkins himself wrote as though there were two alternative natural philosophies adopted by modern thinkers at Oxford:

there is not to be wished a more generall liberty in point of judgment or debate, then what is here allowed. So that there is scarce any Hypothesis, which hath been formerly or lately entertained by Judicious men, and seems to have in it any clearenesse or consistency, but hath here its strenuous Assertours, as the Atomicall and Magneticall in Philosophy, the Copernican in Astronomy &c.

S. Ward (1654), preface by Wilkins, p. 2.

The survival of an alternative to the mechanical (or 'Atomicall') philosophy was significant. The magnetical philosophy, or residual influences stemming from it, allowed a greater conceptual freedom among English cosmologists and a richer theoretical vocabulary. Thus they could formulate theories in terms of forces acting between heavenly bodies. If pressed, they might well appeal to some underlying mechanical account of the apparent influence; but in a more thoroughgoing mechanical philosophy, the freedom to employ concepts of attractive force had been lost. Restrictions on the notion of material action inhibited theories involving action at a distance, even if the action was only apparent.

If we take the case of immediate concern, namely planetary motion, the Cartesian explanation was thoroughly mechanical. Inertial motion, according to Descartes's *Principles of philosophy*, is rectilinear, so that a body constrained to move in a circle experiences a centrifugal impulse. However, the universe, for Descartes, was a plenum – full of matter – so that in a whirling vortex, such as the solar system, the centrifugal tendency of each material particle could be fulfilled only in relation to an overall equilibrium. The system was therefore governed by competing centrifugal tendencies, consistent with a strictly mechanical understanding of material action.

The analysis provided by Descartes thus concentrated attention on the centrifugal tendency of a body constrained to move in a circle, and this was characteristic of later attempts to mathematize the Cartesian account. The magnetical philosophy, however, naturally presented a different approach to the problem, and the physical basis of a more realistic dynamical analysis. It suggested a central attractive force – a centripetal force – continuously drawing the planet from its inertial path.

Through the mediation of a working tradition of cosmological speculation and experimental natural philosophy in England, the influence of Gilbert extended clearly to Wren. In his inaugural address at Gresham, Wren was expanding on the theme of the new liberty in natural philosophy, represented by the triumph of the Copernican theory, when he added a significant appraisal of Gilbert's work, and an important statement of the received attitude to him half a century after the *De magnete:*

Among the honourable Assertors of this Liberty, I must reckon *Gilbert,* who having found an admirable Correspondence between his *Terella,* and the great *Magnet* of the Earth, thought, this Way, to determine this great Question, and spent his Studies and Estate upon this Enquiry; by which *obiter,* he found out many admirable magnetical Experiments: This Man would I have adored, not only as the sole Inventor of Magneticks, a new Science to be added to the Bulk of Learning, but as the Father of the new Philosophy; *Cartesius* being but a Builder upon his Experiments. This Person I would have recommended to Posterity in a Statue, that the deserv'd Marble of *Harvey* might not stand to future Ages, without a Marble Companion of his own Profession. He kept Correspondence with the *Lyncei academici,* at *Rome,* especially with *Franciscus Sagredus,* one of the Interlocutors in the Dialogues of *Gallilaeus,* who labour'd to prove the Motion of the Earth, negatively, by taking off Objections, but *Gilbert* positively; the one hath given us an exact Account of the Motion of Gravity upon the Earth; the other of the secret, and more obscure Motion of Attraction and magnetical Direction in the Earth; the one I must reverence for giving Occasion to *Kepler* (as he himself confesses) of introducing Magneticks into the Motions of the Heavens, and consequently of building the elliptical Astronomy; the other of his perfecting the great Invention of Telescopes, to confirm this Astronomy; so that if one be the *Brutus* of Liberty restor'd to Philosophy, certainly the other must be the *Collatinus*.[4]

Wren sees Gilbert, rather than Descartes, as 'the Father of the new Philosophy', but perhaps more significant is the contrast he draws between the roles played by Gilbert and Galileo. Thus Galileo worked in a negative way to remove difficulties and anomalies in the Copernican theory, but Gilbert brought to the theory new proofs and opened up its development in new directions. In particular he was the inspiration for Kepler 'introducing Magneticks into the Motions of the Heavens, and consequently . . . building the elliptical Astronomy'.

Wren's interest in the elliptical astronomy was growing at around this time. We have seen already that he said in the same speech that 'of all the Arguments which the Learned of this inquisitive Age have busy'd themselves with, the Perfection of these two, Dioptricks, and the Elliptical Astronomy, seem most worthy our Enquiry' (C. Wren, 1750, p. 204). He would have been instructed in Keplerian planetary theory by Seth Ward,[5] but we know that by 1658 he had a first-hand knowledge of Kepler's work, and it must have been around the same time that he delivered his Gresham lectures on the same subject (see C. Wren, 1750, p. 239).

Kepler's so-called 'second law' governed the speed of a planet in its elliptical orbit. In its most correct form, it stated that the straight line joining the planet to the Sun, at one focus of the ellipse, swept out equal areas in equal times. Ward employed a more easily applied approximation according to which the other focus of the ellipse – the 'empty' focus – is an equant point: the straight line joining the planet to the empty focus moves with uniform angular velocity (see Russell, 1964, pp. 17–19). Now, in 1658, Wren was concerned with these related problems, and published – this was quite unusual – a statement of the correct 'area' version of Kepler's second law.

Wren had solved a mathematical problem, posed in France under the pseudonym of Jean de Montfert as a challenge to the English mathematicians (see A. R. Hall, 1965, pp. 140–4). He thought the problem derived from an equant version of the second law and, along with his solution on a printed fly-sheet, presented 'Kepler's problem' as a counter-challenge. This concerned the application of the area law to astronomical calculation and Wren had found it posed in the *Astronomia nova*. His own geometrical solution was published by John Wallis the following year.[6] Wren had given it to Wallis in July 1658 (see Wallis, 1659, p. 70), here Wren repeats the area law, using a different form of words, and is clear that Kepler's account derived *ex causis physicis* (see Wallis, 1659, p. 80 (1st pagination)).

To follow Wren's own understanding of the physical causes, we can return at last to his reply to Halley's question on this very subject, posed at Newton's request. Halley reported to Newton in June 1686:

> According to your desire in your former, I waited upon Sr Christopher Wren, to inquire of him, if he had the first notion of the reciprocall duplicate proportion from Mr Hook, his answer was, that he himself very many years since had had his thoughts upon making out the Planets motions by a composition of a Descent towards the sun, & an imprest motion; but that at length he gave over, not finding the means of doing it. Since which time Mr Hook had frequently told him that he had done it, and attempted to make it out to him, but that he never satisfied him, that his demonstrations were cogent.
> H. W. Turnbull *et al* (1960), vol. 2, pp. 441–2

Wren, it seems, left aside the question of the mathematical relation between gravitation force and distance, but said that, many years previously, he had considered the notion of deriving planetary orbits by combining central and tangential components, and he firmly claimed priority over Hooke.

We cannot say with any certainty how early Wren had reached this fundamental insight, but the form of words used in Halley's report may – just possibly – indicate one of his sources.

In the fourth day of the *Discorsi* Galileo presents a cosmogony which he attributed to Plato and had already explained in the *Dialogo*, according to which God had allowed all the planets to 'fall' towards the Sun, beginning from the same point in the universe and each moving with uniformly accelerated motion.[7] When each planet had achieved its ordained speed, God had converted its rectilinear motion into a uniform circular one. Galileo suggested that calculations based on existing planetary orbits would confirm the possible existence of this unique starting point.

Wren's copy of the *Discorsi* has survived in the Bodleian Library and a Latin annotation in the margin, apparently in his hand, makes a comment on this suggestion:

> If this Platonic hypothesis was true, the Earth and the other planets should have respect to some centre in the sky (perhaps the Sun) as regards their straight motion: (just as our heavy bodies are carried towards the centre of the Earth) or else some other cause of the accelerated motion is to be assigned.
> Bennett (1975), p. 38.

This brings to mind Halley's report, where he spoke of 'a composition of a Descent towards the sun, & an imprest motion'. Hooke, for reasons we will see shortly, would have reversed this order and spoken of 'the inflection of a direct motion into a curve'.

It is very likely that the annotations in this book date from between 1656 and 1660 (see Bennett, 1975, p. 38). Combined with Wren's interest in 'magneticks' as a conceptual tool in cosmology, this hint from Galileo may have helped him to grasp the fundamental dynamical components of orbital motion – 'a composition of a Descent towards the sun, & an imprest motion'. Galileo had assumed a circular inertia: the planet would naturally continue in its circular path, influenced by no other force. But Wren seems to have realized that the central force must continue to act to maintain the planet's orbit. By 1661, Wren had made an experimental study of the laws of impact (see Chapter 7) and must at least by then have been familiar with the mechanics of Descartes's *Principles of philosophy*, with its rectilinear inertia and its analysis of constrained circular motion.[8] Once he had grasped this analysis, the magnetical philosophy would have provided him with the concept of an attractive centripetal force as the constraining principle.

Unlike that of Wren, we can date the beginning of Hooke's interest in the problem with some certainty, and from then on the parts played by the two are impossible to separate. Hooke first announced his programme for a 'System of the world' in a famous statement published in 1674,[9] so often quoted as a striking anticipation of the principles of the Newtonian cosmology. Hooke postulated a gravitational interaction between heavenly bodies, a principle of rectilinear inertia and a decrease in gravitation as some function of distance. What this function was, Hooke said he had 'not yet experimentally verified', and we can trace his experiments back a number of years. They involved weighing bodies at different distances above and below the Earth's surface, and they were relevant to the 1674 programme because, by then, Hooke was identifying gravity with the mutual attractions of heavenly bodies.

This was not true of the various experiments conducted by Hooke between 1662 and 1664, mainly at the suggestion of the Royal Society (for details, see Bennett, 1975, pp. 40–2). They derived from reports, recounted by Bacon in *Sylva sylvarum* and related simply to variations in terrestrial gravity with distance. However, as soon as the Royal Society began to meet once again, following a recess during the Plague, when Hooke had conducted related experiments with Wilkins, he reported the results within a greatly expanded context. He began, on 21 March 1665/6, to give his experiments on the variation of gravity and the nature of gravity a cosmic significance (see T. Birch, 1756, vol. 2, p. 70; note also p. 65).

After making obscure hints at the potential of his work, Hooke embarked on a series of experiments to compare the variations of gravitational and magnetical attractions with distance, but he was prompted to be more specific about his programme by a paper read to the Society by John Wallis on 16 May (see T. Birch, 1756, vol. 2, p. 89; note also p. 88).

Wallis was trying to complete Galileo's explanation of the tides, namely, that they resulted from the motion of the Earth, by adding a further motion to the daily and annual components considered by Galileo. Wallis argued that the primary orbit – 'whether *Circular* or *Elliptical;* of which I am not here to dispute' (Wallis, 1666, p. 272) – was traced out, not by the centre of the Earth, but by the common centre of gravity of the Earth and the Moon. This meant that the Earth performed an epicyclic motion about the common centre of gravity, and the proposal clearly raised other cosmological questions. What was the tie that linked the Earth and the Moon? Wallis chose to remain agnostic, saying that such a tie, whatever its nature, really did exist; however, he naturally reinforced his position by drawing analogies from magnetic attraction (see Wallis, 1666, pp. 271, 282).

At the meeting when Wallis read his paper,

It being mentioned by Mr Hooke, that the motion of the celestial bodies might be represented by pendulums, it was order'd that this should be shewed at the next meeting. T. Birch (1756), vol. 2, p. 90.

So on 23 May, Hooke presented his famous paper 'concerning the inflection of a direct motion into a curve by a supervening attractive principle' (T. Birch, 1756, vol. 2, pp. 90–2). He now explicitly stated his explanation of planetary motion – a combination of rectilinear inertial motion and a central attractive force. He illustrated this with a conical pendulum. A pendulum bob, held to one side and given a tangential impulse, could be made to perform roughly elliptical or circular motions. The link with Wallis's paper was that, with a small secondary pendulum attached to the first, both could be made to perform epicyclic motions about the common centre of gravity which, in turn, performed the primary circular or elliptical orbit.

Hooke had now revealed the cosmological significance of his work – a significance entirely absent from his previous series of experiments in August 1664 – and there is no doubt that Wren was involved in this change.

The way in which Hooke mentioned his pendulum demonstration, on the day when Wallis read his paper, suggests that he was bringing forward a relevant experiment with which he was already familiar. Thomas Sprat attributed the demonstration explicitly to Wren, saying that Wren discovered that a pendulum would continue to move either in *Circular,* or *Eliptical* Motions . . . and that by a complication of several *Pendulums* depending one upon another, there might be represented motions like the Planetary *Helical Motions,* or more intricate: And yet that these *Pendulums* would discover without confusion (as the *Planets* do) three or four several *Motions,* acting upon one Body with differing *Periods.*

Sprat (1667), pp. 313–14.

Further, in the priority dispute between Hooke and Huygens, following the publication of *Horologium oscillatorium*, Oldenburg wrote to Huygens on 27 June 1673:

Touchant le pendule circulaire ie le dis encor, tesmoin le Registre de la Soc. Roiale, qu'il y a plusieurs annees, que M. Hook nous en montra icy les proprietez, et mesmes en fit construire des horologes veuës de plusieurs Estrangers. Et M. Wren en auoit desia parlé devant luy a quelques vns de ses amis icy, qui sont prests d'en rendre tesmoignage.

Huygens (1888), vol. 7, p. 323.

Hooke dated his interest in the conical pendulum to 1665 (see Gunther, 1920, vol. 8, p. 105) and, in using it to demonstrate the nature of planetary motion, it seems he was repeating an illustration he had derived from Wren. But it goes further back, at least in part, for in 1638 Jeremiah Horrox used the conical pendulum as a model of elliptical planetary motion (see Armitage, 1950, p. 278). Horrox's manuscripts excited the interest of the Royal Society and, in 1664, Wallis began to examine them at Oxford with a view to publication.[10] Wren too was involved in the early stages of this work.[11] The reference to the conical pendulum model occurs in a letter from Horrox to Crabtree, which Wallis received in August 1664.[12] This was after he had reported that he and Wren had examined the papers, but it is quite likely that Wren would have seen the later additions as well. In fact, as we shall see shortly, at a slightly later date we can link Wallis, Wren and the relevant letter.

We know that Wren had already discussed the conical pendulum with Huygens, and he probably had a fair grasp of the dynamics involved, even if Horrox was responsible for its explicit use as a planetary model. It was Wren alone, according to Sprat, who developed the model by adding a secondary pendulum illustrating the motion of the common centre of gravity. Other experiments by Wren, involving pendulums, were also relevant, for Wren had shown around 1661 that the motion of the common centre of gravity is unaffected by elastic impact, but we shall deal more specifically with this later. At present we can uncover only the general lines of Wren's interests in the related subjects of mechanics, planetary theory and cosmology. We can, however, discover the likely source of Hooke's involvement in the discussion.

The appearance of a comet was reported to the Royal Society in December 1664, at a meeting at which Hooke and Wren were present.[13] Observations began to come in from England and the Continent and at first Hooke was asked to compile a 'history' (see T. Birch, 1756, vol. 1, p. 511).

At Oxford, both Wallis and Wren were interested. Wallis first observed the comet on 23 December (see A. R. Hall & M. B. Hall, 1965, vol. 2, p. 339). During January, the Royal Society began to look to Wren rather than Hooke for their contribution to the growing discussion, Moray writing to Huygens on 20 January that Wren had all the submitted accounts (see Huygens, 1888, vol. 5, p. 212).

The following day, Wallis reported to the Society on Adrien Auzout's *L' éphéméride du comète . . . fait le 2 Janvier 1665.*[14] Auzout favoured a rectilinear path and Wallis pointed out that, as a general hypothesis on cometary paths, this was not original. It had in fact been suggested by Kepler; but the theorist of special interest to Wallis, at the time, was Horrox, and he went on to reveal some of the ideas he had found in the letters to Crabtree:[15] A comet is emitted from the Sun in a straight line, but is retarded by the attractive force of the Sun and eventually drawn back again. Because of the Sun's rotation, sweeping its magnetic influence round in the manner proposed by Kepler, the resultant path is not straight, and Horrox drew the path of the 1577 comet, observed by Tycho, as an ellipse which began and ended in the Sun.

Wren too had recently been interested in the possible rectilinear motion of comets and, early in January, he had proposed a directly relevant problem to Wallis.[16] It concerned locating a comet's position in space. Parallax measurements were one solution, but Hooke tells us that they were associated with difficulties, inaccuracies and disagreements (see Gunther, 1920, vol. 8, pp. 236–40). Suppose, however, the comet's motion is uniform and rectilinear. The problem of finding the projection of its path on to the ecliptic, given four observations of right ascension, is that of finding the straight line cut by four given straight lines, so that its three portions are related in a given ratio. The ratio is that of the time intervals between the observations. This was the geometrical problem that Wren proposed to Wallis and, much later, in 1677, Wallis suggested using his solution as a possible appendix to an edition of Horrox's works, saying that the problem had been 'proposed to me by Dr. Wren, to be effected in order to find the distance of comets' (Rigaud, 1841, vol. 2, p. 605).

Evidence that Wallis and Wren discussed the possible rectilinear motion of comets is particularly interesting, because the letter to Crabtree (25 July 1638) where Horrox first made the proposal – one of the letters Wallis referred to in his report to the Royal Society[17] – also contained his conical pendulum model of elliptical planetary motion (see Wallis, 1678, pp. 309–14).

On 27 January, Moray again promised Huygens that Wren would soon produce some account from the submitted observations (see Huygens, 1888, vol. 5, p. 228), and at a Royal Society meeting on 1 February; 'Dr. Wren produced some observations of the comet, with a theory' (T. Birch, 1756, vol. 2, p. 12). The theory, it seems, was Wren's method of locating a comet, assuming a uniform rectilinear path. It included Wren's own solution to the geometrical problem he had proposed to Wallis, and an application of his method to the comet in question. This was a diagram, in which Wren used his construction to locate the comet and describe its motion in a straight line between 20 October 1664 and 20 January 1664/5. Hooke was shortly in possession of both, and he later published them in his *Cometa*. The original diagram has been preserved at All Souls College, Oxford.[18]

The success of Wren's construction, in producing the same line for different sets of observations, was a test of the hypothesis of uniform rectilinear motion. Though far from being exactly right, the result seems to have lent it some support, and Hooke said that it 'did very near solve all the appearances preceding and subsequent' (Gunther, 1920, vol. 8, p. 258). Wren was not satisfied. Observations continued to be referred to him and, during February, Moray declined to send details of Wren's work to Huygens, but promised a complete account shortly and said that Wren intended to publish his results.[19] A tract entitled 'De natura & motibus cometarum' was extant after Wren's death, but never published.[20]

In fact, Wren's interest soon began to fade. By late February, Hooke was lecturing on the comet at Gresham College (see Latham & Matthews, 1972, vol. 6, p. 48) and, in March, Wren was engaged in planning a visit to France. The Society turned once again to Hooke, and on 17 March Moray told Huygens that Hooke had taken over the task from Wren (see Huygens, 1888, vol. 5, p. 286).

In his reply of 31 March, Huygens said he had heard of a new comet (see Huygens, 1888, vol. 5, p. 320), and this appearance changed the situation once again. Hooke had already told Moray that he had seen it (see Huygens, 1888, vol. 5, p. 322), and the fresh possibilities immediately re-awakened Wren's interest. Both Hooke and Wren were now committed to the problem, and Wren was occupied with the new comet during the first weeks of April. He wrote to Moray on 11 April that he had made his first accurate sighting on 7 April, and that: 'Since this I have rose every morning but though the dayes be fair & the nights, yet the mornings with us are misty towards the Horizon' (Royal Society MS EL. W. 3 no. 5). This letter shows Wren's revived enthusiasm and the issues raised by the new comet:

I shall take it for a great Favour, if I may obtain what observations you have at Gresham of this 2^d comet. Alsoe I would desire of M! Hooke he would let me have all the last observations of the last Comet when he began to be stationary or slow in his Motion, & when & where he disappeared. I hope my quiet in this place may continue till I have perfected the Hypothesis I began & made a good essay towardes of the last Comet: & I have a great desire to find whither this be not yet the same, for who knowes what disposition of the matter makes the various intention or remission of Light in Comets, & though this last appearance were brighter & more silver coloured then ever the first was, yet as long as I see it in the same path & Retrograde when the other should be retrograde, I have some suspicions it may be the same. is it thus, or else doe comets kindle one an other, or propagate by a kind of Generation? but I suspend till I have more observations.

Royal Society MS EL. W. 3 no. 5.

The question of the relation between the two appearances explains why Wren asked the Royal Society to return his diagram of the path of the first comet. On 19 April, Hooke was asked to copy Wren's diagram and return the original (see T. Birch, 1756, vol. 2, p. 32), the original being, in all probability, the drawing in the collection at All Souls.

The following day, Wren wrote to Hooke:

I thanke you for the freedome of your converse wch. I should be glad you would sometimes continue to me whilest I am heer though I dare not importune you to it, for I know you are full of employment for the Society wch. you allmost wholy preserve together by your own constant paines. I have not yet received the Globe & papers . . . I shall be attent to looke for both the Comets if the Sun give us leave, though I am affrayd the 2d. runnes too fast into the South. The Hypothesis of Cassini[21] of the Comets motion about the Dog starre I can by noe meanes approve. I should rather take Lyra for the center then Syrius, though I am not fond of this neither. I have I thinke lighted upon the trew Hypothesis wch. when it is riper & confirmed by your observations I shall send you.

Royal Society MS EL. W. 3 no. 6.

It would seem that Wren was not bound by the hypothesis of uniform rectilinear motion, and we learn a little more of the exchanges with Hooke from Hooke's reply of 4 May. An edited version was printed in *Parentalia:*

I hope you received the Globe and Observations which I sent you; you had had them much sooner, but in Truth I could not get the Copy of your *Hypothesis,* though the *Amanuensis* was ordered by the *Society* to have had it ready above a Week before. Those Observations of my own making, I have not yet had Time to adjust so well as I desired.

C. Wren (1750), p. 219.

Hooke expected that, in two weeks' time, when the first comet had passed the Sun, he would be able to see both the first comet with a telescope and the second with the naked eye,

for, if it [the first] continue to move those Ways I have imagined it, whether we take the Supposition of the Motion of the Earth, and imagine the *Comet* to be moved in a Circle . . . or whether we suppose the Earth to stand still, and the *Comet* to be moved in a great Circle whose convex Side is turned towards the Earth . . . it must appear again very near the same Place about a Fortnight hence. And I am apt to think the Body of the *Comet* is of a Constitution that will last much longer than either a Month or a Year, nay than an Age; and if I can be so lucky to meet with it again, I hope to trace it to its second appearing.

C. Wren (1750), p. 220.

Unfortunately at this point Christopher Wren Jnr (editor of *Parentalia*) omits a section of the letter, so that we know only that Hooke concludes:

But I weary you with my Conjectures; and I doubt not but that before this, you have perfected the *Theory of Comets,* so as to be able to predict much more certainly what we are to expect of these *Comets* for the future; whereof if at your Leisure you will please to afford me a Word or two, you will much oblige me.

In spite of encouragements from the Royal Society, there is no further record of any revised solution by Wren. What is clear is that Hooke and Wren engaged in a free exchange of ideas on the nature and motion of comets, and Hooke's contemporary lectures on the subject show that the comets prompted significant questions for cosmology.

A large section of Hooke's *Cometa,* published after the appearance of a comet in 1677, was taken from various lectures and papers of 1665-6 (see Gunther, 1920, vol. 8, p. 260). It consists of an edited collection of passages drawn from various sources into a less than consistent whole (see Gunther, 1920, vol. 8, pp. 223–60), and the sources must range from his Gresham lectures of 1665 to the paper that Hooke eventually presented to a Royal Society meeting on 8 August 1666, when discussion of the conical pendulum model of planetary motion was continuing (see T. Birch, 1756, vol. 2, p. 107). The period covered is, therefore, one crucial to the development of Hooke's ideas and a careful study shows that at an early date, around March 1664/5, Hooke was thinking in terms of circular paths (compare his letter to Wren quoted above), but that later he settled for paths that were basically rectilinear (inertial), but were modified by gravitational influences (see Bennett, 1975, p. 57).

The details of Hooke's ideas are not of immediate concern here, but it is clear that discussion of the modification of the rectilinear path to fit the observations was an important step. Within the magnetical philosophy, the source of such modification was clear and when Hooke spoke of the 'sphere of activity' of a planet, he echoed Wilkins's 'sphere of vigor' and Gilbert's *'orbis virtutis'*. It was the success of Wren's theory of rectilinear motion that changed Hooke's earlier idea of circular paths, but its failure, in turn, was the source of important cosmological questions. The comet posed, for Hooke, in a dramatic way, the dynamical situation of a rectilinear inertial path modified by a central force, and he eventually accounted for it in these terms. It was a small step to the closed orbit of a planet:

And particularly by tracing the way of this Comet of 1664. It is very evident that either the observations are false, or its appearances cannot be solved by that supposition, without supposing the way of it a little incurvated by the attractive power of the Sun, through whose system it was passing, though it were not wholly stayed and circumflected into a Circle.

Gunther (1920), vol. 8, p. 260.

The comets also brought Hooke into relevant discussions with Wren who, as he said, had already considered planetary motion in these terms. They had opportunities to exchange ideas during 1665 and 1666,[22] in spite of Wren's visit to France; and we have already seen Hooke's experiment with the conical pendulum linked with similar work by Wren. Hooke himself dated his interest in the conical pendulum to 1665 (see Gunther, 1920, vol. 8, p. 105).

The cosmological discussions of Wren, Hooke and, to a lesser extent, Wallis, were full of exciting possibilities, but they remained the basis of a future programme, though Hooke's diary shows that he and Wren continued the debate from time to time.[23] They went only one stage further. Huygens, analysing circular motion within a Cartesian framework, published a formula for centrifugal force in 1673. A centrifugal force, proportional to the velocity squared over the radius, could be immediately interpreted in the magnetical philosophy as a centripetal force which, by substituting for time from Kepler's third law, was found to be inversely proportional to the square of the radius. It may, again, have been Wren who took this step (see Lawrence & Molland, 1970). The result, thus far, applied only to a circular orbit.

There it rested until Hooke roused Newton to carry the matter further and thus to realize its full potential. But he did more than merely re-awaken Newton's interest. For until Hooke's intervention, Newton, schooled on the books of the mechanical philosophy, and especially influenced by Descartes, had, like Huygens, analysed the dynamics of circular motion in terms of a constrained centrifugal force (see Whiteside, 1964 and 1970). Hooke, from a different natural philosophical standpoint, turned the problem around and presented him with a centripetal force continuously modifying an inertial path. Many years of discussion and development – a development in which Wren had played a crucial role – were about to reach fruition.

7

Mechanics, microscopy, surveying

Fundamental to any application of Cartesian philosophy in explaining the natural world were the laws of impact contained in Descartes's *Principles of philosophy*. Since action was restricted to collision, these laws governed all physical phenomena, and they became a focus for critical study of Descartes's mechanics.

When Huygens was staying in London in 1661, he was visited on 23 April by a group of English mathematicians – Moray, Brouncker, Paul Neile, Wallis, Rooke, Wren and Goddard – and he recorded in his journal, 'Resolus les cas qu'ils me proposerent touchant les recontres de deux spheres' (Huygens, 1888, vol. 22, p. 573). Prompted by Oldenburg in 1665, Moray recalled the occasion, 'as both Dr. Wallis and I do well remember'.[1] Huygens, apparently, had raised the subject of impact and since Wren and his colleague at Gresham College, Laurence Rooke, had already made some experiments – 'with balls of wood & other stuff hanging by threads' – they proposed a number of initial conditions to Huygens. Their host used his rules, which he did not reveal, to derive the results that Rooke and Wren had already achieved by experiment.

It was probably in 1661, though after the discussion with Huygens, that Wren evolved his own theory of elastic impact, a theory which came to light when the Royal Society took up the question in 1668. When the theories of Huygens and Wren were published in 1669, it was clear that they differed only in formulation.[2]

We can date Wren's theory fairly closely. Huygens said that, while Rooke and Wren had made experiments before April 1661, they had not reached a theory (see Huygens, 1888 vol. 6, pp. 383, 386). When Wren, prompted by Oldenburg, presented his theory at a Royal Society meeting on 17 December 1668, he said that,

he had this hypothesis several years before, when the society began to be formed; and that Mr. Rooke and himself made divers experiments before the society to verify the same: which affirmation of his was seconded and confirmed by several of the members, who were eye-witnesses of those experiments, as the president [Brouncker], Sir Paul Neile, Mr. Balle, and Mr. Hill.[3]

Sprat, in his history of 1667, mentioned Wren's experiments – 'produc'd before the Society' – as well as his theory.[4] The Society was formed towards the end of 1660 and, since Rooke died in 1662, it seems likely that Wren's theory dates from 1661. In January 1661/2, Wren was also interested in the mechanics of free fall.[5]

Wren's theory – 'Lex Naturae de Collisione Corporum' – employs a somewhat unusual formalism. He begins by defining what he means by 'proper' velocities of bodies in collision: 'The proper and most truly natural velocities of bodies are reciprocally proportional to the bodies' (A. R. Hall & M. B. Hall, 1965, vol. 5, p. 320). The theory states that if two bodies move with proper velocities before collision, that is, if their velocities are inversely proportional to their respective bodies (we would say 'masses'), their speeds are unchanged after collision, but their directions are reversed. Wren illustrated this case by a balance in equilibrium, where the weights correspond to the two 'bodies' and their distances from the fulcrum correspond to the respective velocities. If the velocities are 'proper', the collision is 'balanced':

For this reason the collision of bodies having their proper velocities is equivalent to a balance swinging about its centre of gravity.
                              A. R. Hall & M. B. Hall (1965), vol. 5, p. 320.

Using the balance model, the case of proper velocities is that of the fulcrum at the centre of gravity. With improper velocities, that is, when the velocities before collision are not inversely proportional to the bodies, the fulcrum (determined by the velocities) is displaced to one side of the centre of gravity (determined by the 'bodies'). Wren's theory then states that the situation after collision will be represented by the fulcrum displaced an equal distance on the other side of the centre of gravity.

Wren's curious formalism seems designed to bring out the symmetry of the situations before and after collision. It is presented in a rather bald manner – there is no derivation or discussion. It appears as a mere synthesis of experimental results, constrained in its expression by certain unspoken regulative principles of simplicity and symmetry.

The question of a demonstration or derivation of the theory is not an artificial one, for its absence was a weakness obvious to contemporary critics. As soon as Huygens, for example, had received Wren's theory and realized it was equivalent to his own, he was anxious to hear Wren's answer to this question:

I very much hope to learn whether Mr. Wren has also looked for some demonstration of them [the laws of impact] and to see what method he has made use of for that end, or else whether he has only established the law of nature which he proposes in this subject on the basis of experiment.

A. R. Hall & M. B. Hall (1965), vol. 5, p. 362.

We will come back to Wren's answer later, in trying to characterize his attitude to mathematical theory and the metaphysics that might be invoked to justify it.

Microscopy is another area where Wren took an early step in a study that came to have the greatest importance. In 1665 Nicholas Mercator told Samuel Hartlib that Wren was making improvements in microscopes, 'to make them multiply exceedingly and geometrically to measure the things in them' (G. H. Turnbull, 1952, p. 112); and in September of the same year, Wren told Hartlib that he was preparing a book with plates of microscopical observations (see G. H. Turnbull, 1952, p. 114). The following year, Wren claimed that his improvements to the microscope, as well as to the telescope, were founded on a theory of refraction (see Bennett, 1973, pp. 146–7). In a reference of 1659 to Wren's work, his cousin Matthew Wren, another member of the Oxford group, stressed, as Mercator had done, its quantitative aspect (see M. Wren, 1659, preface). It may be that Wren was trying to apply a form of micrometer to the microscope, as well as to the telescope.

A letter written by Wren in 1652 shows that he was probably at work on microscope illustrations even then,[6] and this and other references indicate that Wilkins also was interested.[7] It may, in fact, have been Wilkins who suggested Wren's project, for he had written in 1649:

there are many common things of excellent beauty, which for their littleness do not fall under our sence; they that have experimented the use of Microscopes, can tell, how in the parts of the most minute creatures, there may be discerned such gildings and embroderies, and such curious varietie as another would scarce believe.[8]

Wren presented some of his drawings to Charles II, probably early in 1661.[9] Both Huygens (see Huygens, 1888, vol. 3, p. 296; note also p. 286) and Monconys (see Monconys, 1665, vol. 2, p. 82) saw them. Monconys mentioned drawings of a louse, a flea, and the wing of a fly – the kind of drawings that became famous through Hooke's *Micrographia* of 1665. In spite of Royal Society requests, Wren declined to pursue his work further and the Society turned instead to Hooke.[10] He satisfied himself that Wren had left the field, before beginning the work that was to culminate with the *Micrographia* (see Hooke, 1665, preface).

There is an interesting link between Wren's work on the theory of impact and that on microscopy, for both were, at least in part, stimulated by the rising mechanical philosophy. The laws of impact governed the mediation of all influences in the natural world. Microscopy was an allied study, for it held out the hope of basing theories about the minute particles involved on experiment and observation.

Elements of this notion can be found in Descartes's 'Dioptrique' (see Descartes, 1965, p. 172), while Henry Power, in 1664, and Hooke, in 1665, are quite explicit that the microscope is the key to placing the mechanical philosophy on an empirical basis.[11] Wren too shared something of this hope, for he said in 1657:

natural Philosophy having of late been order'd into a geometrical Way of reasoning from ocular Experiment, that it might prove a real Science of Nature, not an Hypothesis of what Nature might be, the Perfection of Telescopes, and Microscopes, by which our Sense is so infinitely advanc'd, seems to be the only Way to penetrate into the most hidden Parts of Nature, and to make the most of the Creation.[12]

Matthew Wren wrote, in 1659, of the Oxford group:

They have sometimes had occasion to inquire, by the help of a Microscope, into the Figure and position of those smaller parts of which all Bodies are composed.

M. Wren (1659), preface.

It was the emphasis given in the mechanical philosophy, to the minute particles of matter, that encouraged the development of microscopy in the mid-seventeenth century. Though invented at around the same time as the telescope, the microscope had not had a similar impact because results of observations were not then relevant to any important theoretical debate.

Another of Wren's less significant interests – surveying – can hardly be seen in relation to any theoretical context, but is interesting rather as one of the traditional mathematical sciences. It was, in fact, quite an early interest, for the *Parentalia* list of Wren's projects at Wadham includes: 'To Measure the straight Distance, by travelling the winding Way'; 'To Measure the Basis and Height of a Mountain, only by journeying over it'; 'A Perspective Box, to survey with it'; and 'A Scenographical Instrument, to survey at one Station' (C. Wren, 1750, pp. 198–9).

Of these schemes – each of which was associated with an instrument – only the 'Scenographical Instrument' is well documented. Its purpose was to enable an unskilled hand to draw any object in perspective and, as early as 1653, the instrument-maker Ralph Greatorex reported a related invention of Wren's to Hartlib (see G. H. Turnbull, 1952, p. 115). Wren himself described it to him in 1655 (see G. H. Turnbull, 1952, p. 115).

It is not clear how, some years later, such an instrument came into Oldenburg's possession, but he showed one to Monconys in June 1663 (see Monconys, 1665, vol. 2, p. 62). It had been made by Anthony Thompson,[13] and Monconys was so pleased with it (see Sorbière, 1709, pp. 27–9) that he wanted to order one himself.[14] It was also shown to Prince Rupert in November 1663 (see T. Birch, 1756, vol. 1, p. 329). Pepys saw the instrument at Oldenburg's in April 1669, and had one of his own made.[15] Boyle showed it to Cosimo III of Tuscany the following month (see Maddison, 1951, pp. 26–8), and Grew records an example in the Royal Society's museum.[16]

It was in March 1669 that the design reached a wider audience, for Oldenburg published, in the *Philosophical Transactions*, 'The Description Of an Instrument invented divers years ago by Dr. Christopher Wren, for drawing the Out-lines of any Object in Perspective' (*Philosophical Transactions*, 1669, **4**, 898–9). The description is probably Oldenburg's and his diagram (see Figure 4) corresponds with that recorded by Monconys in 1663. The instrument itself is simple. The object is viewed through a pinhole sight and its outlines traced by a pin (at an adjustable distance from the sight) attached to a horizontal ruler. By a system of

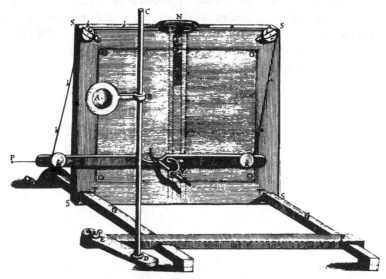

Fig. 4. Wren's perspectograph. (From *Philosophical Transactions*, 25 March 1669, vol. 4,, no. 45, facing p. 893.)

pulleys and weights the ruler is counterbalanced and kept horizontal and a pen, near its centre, is pressed by a spring onto paper fastened to a board. The outline traced is thus transferred to the paper.

Of greater use to surveying proper was the only other relevant instrument by Wren of which we have any details. Once again, the published account was due to someone else, but on this occasion it was Hooke. In 1666, prompted, no doubt, by the surveying activity following the Great Fire of London, Wren designed 'a new level for taking the horizon every way in a circle'.[17] Wren's own description has survived, and Hooke – one of the surveyors specially appointed after the Fire – published an account (see Gunther, 1920, vol. 8, p. 101). The level consisted of a large, shallow concave of glass, mounted on a ball and socket joint. Its accurately ground edge lay horizontal when a drop of mercury rested precisely in the centre, '& consequently a Dioptra layd upon it will giue an exact Levell in any Azimuth without motion of the Instrument' (Royal Society MS CP. II. 1); Hooke had used an instrument of this kind.

Mechanics, microscopy and surveying are some examples of Wren's minor interests. Surveying was, of course, one of the traditional mathematical sciences. Wren's interest in mechanics and microscopy, on the other hand, were stimulated by the contemporary mechanical philosophy. Yet, typically of Wren, instrumentation is common to them all for, even in mechanics, Sprat speaks of his 'Instrument to represent the effects of all sorts of Impulses, made between two hard globous Bodies' (Sprat, 1667, p. 312). The mathematical sciences were an important influence on natural philosophy – the mechanical philosophy in particular – and Hooke and Wren are examples of their fruitful interaction. The interaction is also seen when we look at Wren's biomedical interests.

8

Medicine and meteorology

Two aspects of Wren's interest in the biomedical sciences – his experiments on intravenous injection and his collaboration with Thomas Willis on the anatomy of the brain – are already well known. Their particular fame is explained by subsequent developments in medicine, but other work was no less important to Wren.

The conceptual background to Wren's injection experiments lies in William Harvey's theory of the blood's circulation, and his most direct link with Harvey was Charles Scarburgh.[1] Scarburgh had known and worked with Harvey in Oxford during the 1640s. The immediate stimulus, however, may have been the injection of poisons in nature. Dean Wren, in an extended note to Browne's *Pseudodoxia epidemica*, discussed the poisonous bite of the scorpion. The note shows how reflecting on this natural injection might lead to the idea of an artificial one. Its last paragraph reads:

Oyle, by nature, abates & duls, & retundes ye feircenes, & spreading of Poyson injected, into vs, by venimous creatures, where wee may come to applye itt: But being dull of itt selfe, & not able to follow ye swift spreading of the Scorpions Poyson, through soe small a Puncture, as ye Scorpion makes; Itt was happilye discouered, that ye Spirit of ye dying Scorpion, receiued in oyle Oliue, (& wth itt chafte upon ye Puncture, as soone as itt is felt;) followes ye Poyson injected, by ye same waye; & soe making way for ye Oyle, wherein itt is caryed, caryes ye Balme, that kils & deades, ye Killing Poyson, before itt can seise, on our Vitall spirits, to destroy them: And noe doubt, but ye Oyle wherein Hornets are drowned would cure y^{eir} Punctures, alsoe; A thing worthe ye Tryall.

Bodleian Library, 0.2.26 Art. Seld., p. 178.

Robert Boyle tells us that Wren's suggestion of the injection experiment came when he was discussing 'the application of poisons' with Wren and Wilkins (see T. Birch, 1772, vol. 2, p. 88).

Boyle has left us the fullest account (see T. Birch, 1772, vol. 2, pp. 88–9) of Wren's experiments of around 1656, in which a variety of substances were injected into dogs. Wren himself, in a letter to Petty, mentions a number of experiments with different infusions, but says only that the work will be 'of great concernment, and what will give great light both to the Theory and Practice of Physick'.[2]

There is evidence that Wren studied arterial structure and action through both dissection and vivisection (see T. Birch, 1756, vol. 4, p. 120), and his interest in Harveian physiology extended also to generation. In 1662 he urged on the Royal Society, programmes of research into generation in hens and rabbits, in particular to test some of Harvey's ideas.[3] Wren himself, in 1663, following the example of Harvey (see Clark, 1898, vol. 1, p. 300), studied the hatching of eggs, his work in this instance being an application of his design for a constant-temperature furnace.[4]

The fact that Wren recorded a technique he used for removing the spleen from a live dog (see C. Wren, 1750, pp. 237–8) illustrates again his concern with current issues in physiology. The function of the spleen was a complex problem under contemporary discussion and splenectomy was in vogue as a relevant experiment.[5]

It was around the same time as his interest in the incubation of eggs, that Wren was working with Willis on the anatomy of the brain and preparing some of the most outstanding of the illustrations in Willis's *Cerebri anatome* of 1664. The subject had interested Wren for some years for, in a report on work done at Oxford around 1656 (the letter to Petty, mentioned above), he says:

Some Parts of Animals we have more exactly trac'd by the help of Glasses, as . . . the Plexus in the Brain. The Nerves we have found to have little Veines & Arteries in them.

                                                                    Bennett (1973), p. 147.

A series of letters from Richard Lower, who collaborated with Willis, to Robert Boyle shows that their work had begun early in 1661, but Wren's contribution is not mentioned until June 1663, by which time his illustrations were almost complete (see T. Birch, 1772, vol. 6, pp. 462–6, 487). Willis's preface, however, implies a more intimate connection with the work, for he says that both Wren and the physician Thomas Millington 'were wont frequently to be present at our Dissections, and to confer and reason about the uses of the Parts'. His tribute to Wren is generous:

Dr Wren, was pleased out of his singular humanity, wherewith he abounds, to delineate with his own most skilful hands many Figures of the Brain and Skull, whereby the work might be more exact.

                                                                    Feindel (1965), vol. 2, preface.

The most sustained of Wren's physiological interests has received least attention, but it too is rightly understood within a collaborative programme at Oxford. This was not a joint investigation, such as the study of the cerebral anatomy but, rather, a web of mutually discussed and commonly held interests and theoretical ideas. It concerns the related topics of muscular action and respiration.

Wren's interest in muscular action has been linked in the secondary accounts with the known expertise of Charles Scarburgh. On the basis of some passages in *Parentalia* it is generally said that before he went to Oxford as an undergraduate, Wren acted as a 'demonstrator' at Scarburgh's lectures on muscles at Surgeons' Hall, and that he made pasteboard models to illustrate the lectures. However, a critical examination of these passages[6] shows that there is no basis for so early a date for Wren's interest in muscular action, nor for the picture of Wren as an assistant to Scarburgh, nor for the idea that he made models for Scarburgh's lectures. Christopher Wren Jnr, when he compiled *Parentalia*, does seem to have had some evidence that Wren made models illustrating muscular action and that he wrote a treatise on the subject, but beyond this his account is very doubtful.

Evidence from other sources dates Wren's interest in muscular action to the early 1660s – a time when he was collaborating at Oxford with Boyle, Lower and Willis, and his ideas fit well with the iatrochemical notions of his associates. They also fit within a whole group of ideas held in common with his friends.

Scarburgh was known for a geometrical and mechanical account of the muscles.[7] As a friend and scientific associate, Wren may well have benefited from discussions with Scarburgh, but Wren was clear that a chemical account was needed to complement the mechanical. He told the Royal Society, in an address which probably dates from the beginning of 1662:

in the Body of a Man, if we consider it only mechanically, we may indeed learn the Fabrick and Action of the organical Parts, but without Chymistry, we shall be at a Loss to know, what Blood, Spirits and Humours are, from the due Temper of which (as of the Spring in the Barrel Wheel) the Motions of all the Parts depend.[8]

The components of the traditional physiology – 'Blood, Spirits and Humours' – must be understood afresh through experimental chemistry, a fundamental tool in advancing beyond the limits of the mechanical philosophy:

Mechanical Philosophy only teaches us what probably may be done in Nature by the Motion and Figures of the little Particles of Things, but Chymistry helps to determine what is actually done by the Motions of those invisible Parts of Liquors, Spirits, and Fumes; and oftentimes gives Light enough to contradict mechanical Hypotheses, that otherwise seem well grounded.

C. Wren (1750), p. 221

The one relevant recorded experiment of Wren's was shown to the Royal Society in 1665: it was proposed at a meeting on 8 March 1664/5 by John Wilkins, who acknowledged it as 'Dr. Wren's suggestion' (T. Birch, 1756, vol. 2, p. 20). A 'fermenting liquor' was to be placed in a jar connected to a bladder; the 'air' being generated could then be collected in the bladder and, by means of a tap, retained. There followed a series of such experiments, conducted before the Royal Society over the next few weeks.[9] At the first experiment, on 15 March, nitric acid was poured on to powdered oyster shells and, 'the exhalation caused by the corrosion of the shells by the aquafortis, in a very little time blew up the bladder . . . so as to swell it with air very plump.' Whereupon: 'Dr. Wren made use of this experiment to explain the motion of muscles by explosion.'[10]

Muscular action by chemical explosion was a notion not unique to Wren. It derived from the not uncommon idea that muscles contract because they are inflated by spirits passing along the nerves. A chemical reaction *within* the muscle seemed, however, a more plausible account of the generated force. In his *De ratione motus musculorum* of 1664, William Croone suggested that the muscular inflation is caused by a chemical reaction between a nervous juice and the blood: 'no one is such a novice in Chemistry as not to know how great a commotion and agitation of the particles is accustomed to occur from different liquors mixed with each other' (Wilson, 1961, p. 162). In the same year, Willis proposed, in the *Cerebri anatome,* a chemical explanation of muscular action. This involved a reaction between the animal spirits carried by the nerves and 'nitrosulphureous particles' supplied by the arterial blood, which produced an expansive force 'like the explosion of Gun-powder'.[11]

These ideas of Croone, Willis and Wren were part of a theoretical web, held in common with other associates and extending to other areas of study. Respiration was directly related through the 'nitrosulphureous' particles carried by the arterial blood. Much of the historical interest in this question has focused on John Mayow, who became a Fellow of All Souls in 1660 and was therefore probably in contact with Wren. Mayow developed a theory (first published in 1668) in which nitrous particles, a vital constituent of the air, were carried by the arterial blood to the muscles and there exploded on contact with the animal spirits. These particles were also essential to combustion. The historical research, stimulated by interest in Mayow's theory, has shown that similar ideas on the vital role of an atmospheric 'nitro-aerial spirit' were held by a good number of his contemporaries, and among Wren's associates these included Willis, Lower, Boyle, Hooke and Ralph Bathurst.[12]

Wren's ideas on respiration fall in with this pattern, since he too held that the air had a vital nitrous constituent. He was particularly interested in demonstrating this by means of what Sprat described as: 'Instruments of Respiration, and for straining the breath from fuliginous vapours, to try whether the breath so purify'd will serve again.'[13] Wren left a written description of such an instrument (see C. Wren, 1750, p. 243), though it seems unlikely that it was ever constructed. But in July 1663 he suggested such a demonstration as entertainment for the King's proposed visit to the Royal Society. His letter to Brouncker reveals the theoretical significance of his instrument:

It would be no unpleasing spectacle to see a man live wthout new Aire, as long as you please. A description of ye vessel for cooling and percolating ye Aire at once I formerly showed ye Society, and left with Mr. Boyle. I suppose it worth putting in practice. You will at least learne thus much from it; if[14] something else in Aire is reqsite for life, yn yt it should be coole only, and free from ye fuliginous vapors and moisture, it was infected wth in exspiraõn; for all these will in probability be separated in ye circulation of ye breath in ye Engine. If Nitrous fumes be found requisite (as I suspect) wayes may phaps be found to supply yt too, by placing some benigne Chymicall Spirits, yt by fuming may impregnate ye Aire within ye vessell.

Royal Society MS EL. W. 3 no. 3.

The reference to a vital nitrous component in the atmosphere, indicates a link between Wren's ideas on respiration and those on muscular action – a link confirmed in Hooke's *Diary*. On 17 December 1677 Hooke noted that Wren 'told me of his paper Mr. Boyle had not returned him, about the fabric of the muscles' (H. W. Robinson & W. Adams, 1935, p. 334). And, on 9 February 1677/8, Wren 'Spake of his Theory of Respiration, muscular motion, &c., delivered to Mr. Boyle.'[15]

Since Wren said, in 1663, that Boyle had his description of a device for purifying the air, since he specified the importance of a nitrous element at the same time and, since he explained 'the motion of muscles by explosion' early in 1665, it seems likely that the theory he mentioned to Hooke dated from the early 1660s and that Wren's ideas developed alongside those of his friends at Oxford. Boyle was doing relevant experiments during this period.

Willis, Lower and Wren were collaborating over the *Cerebri anatome,* and the muscular account it contained was developed by Willis in later publications. Lower linked a nitro-aerial spirit with muscular explosion in his *De corde* of 1669.

In April 1678, only a few months after Hooke and Wren had been discussing Wren's earlier theories, Hooke proposed an explanation of muscular action at a meeting of the Royal Society, with Wren in the chair. When he had finished

an occasion was taken, to discourse of the causes of the motion of the muscles; and how far the air taken in by the lungs might contribute towards muscular motion. And it was thought, that it was of great necessity for that very purpose.

T. Birch (1756), vol. 3, p. 402.

On the question of muscular action, Wren's ideas had not substantially changed:

Sir Christopher Wren supposed, that the swelling and shrinking might proceed from a fermentative motion arising from the mixture of two heterogeneous fluids.[16]

It may be that the story of Cornelius Drebbel's submarine, which leant support to the idea of a vital nitro-aerial spirit because the submariners were revived by breathing an aerial substance prepared from saltpetre, was a stimulus to Wren, as it had been to Boyle and Wilkins.[17] In particular, Wilkins's *Mathematical magick* (1648) influenced Wren in a number of ways and contained a chapter on submarine navigation, discussing 'the greatest difficulty of all . . . how the air may be supplied for respiration' (Wilkins, 1802, vol. 2, pp. 188–94). The *Parentalia* catalogue of Wren's early work contains not only 'Strainer of the Breath, to make the same Air serve in Respiration', but also 'Ways of submarine Navigation' and 'To stay long under Water' (C. Wren, 1750, p. 198).

If the air had such an important role in human physiology, the study of the atmosphere was immediately relevant to health. The pioneering work in meteorology, which was done in England at this time, was seen within the context of an extensive environmental approach to medicine.

As early as 1657, Wren had suggested to the 'rational philosophical Enquirer into Medicine' that a correlative study of dissection, epidemics, the weather and other natural phenomena would yield 'a true Astrology to be found by the enquiring Philosopher, which would be of admiral Use to Physick' (C. Wren, 1750, pp. 202–3). He returned to this theme in his 1662 address to the Royal Society, where he stressed the importance of chemistry to physiology. The programme was now better organized, perhaps because its theoretical rationale – the role of the air in the health of man – had been formulated more clearly:

there is another Part of Physiology, which concerns us as near as the Breath of our Nostrils, and I know not any Thing wherein we may more oblige Posterity, than that which I would now propose.

C. Wren (1750), p. 222.

What Wren proposed was a 'History of Seasons', divided into two parts. The first, 'A meteorological History', consisted of five sub-histories in which were recorded the changing qualities of the air, such as its motion (winds), heat, cold, moisture, or refraction as observed with astronomical instruments. This would be correlated with 'A History of Things depending upon Alteration of the Air and Seasons' – a record of crops and cattle, wines (though, as a foreign import, this belonged rather among the independent variables), fish, fowl, insects and venomous creatures, and

Above all, the Physicians of our Society should be desir'd to give us a good Account of the epidemical Diseases of the Year; Histories of any new Disease that shall happen; Changes of the old; Difference of Operations in Medicine according to Weather and Seasons, both inwardly, and in Wounds: and to this should be added, a due Consideration of the weekly and annual Bills of Mortality in London.[18]

Wren played a significant part in the genesis of interest in the barometer in England, and in the barometer's evolution from the Torricellian tube. The latter was an instrument for demonstrating Torricelli's theory of the weight of air. Its physical arrangement coincided with that of a mercury barometer, but it fulfilled quite a different conceptual function. There was not necessarily any theoretical reason for measuring small variations in the height of the mercury column of a stationary Torricellian tube.

Such measurements were made before Wren suggested a reason, but in England at least, he was understood to have provided the stimulus. Descartes had held that tides on Earth were caused by a mechanical pressure on the water, exerted by the Moon, and transmitted by the aether. When Wren pointed out that careful observation of the Torricellian tube in relation to the lunar motion would test the theory, Boyle took up the suggestion and found a variation unrelated to the passage of the Moon.[19]

Once this variation was seen as an interesting reflection of the state of the atmosphere, the barometer was born, but theories on the cause of 'the alteration of the gravity of the air' – the terms of a Royal Society discussion of November 1679 – were open to debate. On this occasion Wren's ideas reveal the importance of the barometer in the programme of environmental medicine:

Sir Christopher Wren was of the opinion, that it proceeded most of all from the impregnating of the air by nitrous salts, which were continually raised up into it.

T. Birch (1756), vol. 3, p. 509.

Wren was interested in the Torricellian tube in its own right, and in related experiments,[20] and he wrote that 'not every Year will produce such a Master-experiment as the *Torricellian,* and so fruitful of new Experiments as that it . . . ' (C. Wren, 1750, p. 225). Sprat tells us that Wren's instruments for 'finding the gravity of the Atmosphere' included sensitive balances. We have no firm evidence on this,[21] though Monconys says that Wren demonstrated to him a balance, impressive because 'vne quatre-centiesme partie de grain la fait tres bucher'.[22]

Wren also devised instruments for measuring humidity,[23] and a self-registering rain-gauge, designed to empty each time the liquid filled a known volume.[24] Similarly, his interest in the thermometer led to at least two forms that were self-recording.

Strictly speaking, these were not thermometers but 'weather-glasses', for they were not sealed and they registered the combined effects of temperature and pressure changes. One form – probably the earlier – consisted of a large bulb of air, connected to a 'U'-shaped tube containing mercury. A float on the free surface of the mercury was attached to a weight by a string and pulley wheel, and the string could be made to move an index, or to pencil a trace on a sheet moved by clockwork.

This method of registering the changes in mercury level was, of course, applied by Hooke to the barometer, to produce the better-known 'wheel barometer'. Wren's application to the weather-glass was earlier, and the distinction was pointed out explicitly by Robert Boyle when describing his 'new kind of Baroscope' in 1666:

This instrument being accommodated with a light wheel and an index (such as have been applied by the excellent Dr. Christopher Wren to open weather-glasses, and by the ingenious Mr. Hook to baroscopes) may be made to shew much more minute variations than otherwise.[25]

The earliest dated reference occurs in January 1662 when, at a meeting of the Royal Society, according to the Journal: 'Dr. Wren read a paper concerning weather-glasses' (T. Birch, 1756, vol. 1, p. 74). John Evelyn's personal record, however, was that: 'Dr. Wren produced his ingenious Thermometer' (E. S. de Beer, 1955, vol. 3, p. 314). At this point it becomes difficult to distinguish clearly between this first form of weather-glass and a more sophisticated development. Monconys described the former as part of

Wren's weather-clock (see Monconys, 1665, vol. 2, p. 53) which he saw in June 1663, and it is included in an undated drawing by Wren of a weather-clock (see Figure 5).[26]

But Monconys also saw a more subtle device. This was a drum – pivoted horizontally, and acting as the reservoir of air – with a tube leading from it and wrapped around the outside. The tube contained a liquid and had one end open. A change in the volume of the trapped air moved the liquid and altered the centre of gravity of the drum whose motion was registered by an index.[27]

In July 1663, Wren suggested that his 'circular thermometer', as it was sometimes called (see Sprat, 1667, p. 313), might be shown during the King's proposed visit to the Royal Society: 'He mentioned the turning glass thermometer with an index, left with Dr. Goddard.'[28] Wren repeated the suggestion when, as he had promised to do, he wrote to Brouncker on the subject of the visit:

I have pleased myself not a little wth ye play of ye weather wheele (ye only true way to measure expansions of ye Aire).[29]

and went on to suggest a brass pipe containing mercury, whereas Monconys had seen a glass pipe with water.

Wren's 'weather-wheel' was incorporated into a revised weather-clock design, submitted to the Royal Society in December 1663 (see Figure 6).[30]

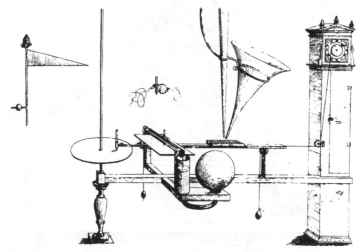

Fig. 5. Design for a weather-clock. (From the 'Heirloom' copy of *Parentalia* at the Royal Institute of British Architects.)

This recorded wind directions and the readings of the weather-glass; and Wren's weather-glass, moved by mercury, is of the second type, though the tube is now concealed within the recording cylinder.

Wren's second weather-clock, omitting the rain-gauge of the first, was a more compact and economical design, and his account shows some attention to practical detail. The recording surfaces, he suggested, might be paper – 'using bread to efface the old tracks' – or boxwood, ivory or silver, 'if the lines engraven be soe filled, that the pencills stick not in them to hinder the Motion' (Royal Society MS EL. W. 3 no. 4).

While the idea of a self-acting weather-clock had interested Wren even before his undergraduate days (see C. Wren, 1750, p. 185) when he first made the suggestion to the Royal Society in January 1661/2,[31] he explained its role within his programme for investigating environmental factors in disease (see C. Wren, 1750, p. 224). The theoretical basis of Wren's biomedical interests was closely related to that of his associates and friends. His particular contributions often derived from his background in the more practical mathematical sciences – he was the draughtsman, the dextrous experimenter, the deviser of instruments – and here he was developing the instrumentation of meteorology as an investigative tool of medicine.

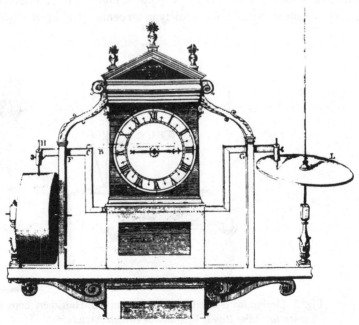

Fig. 6. Design for a weather-clock. (From T. Birch, 1756, vol. 1, plate 3.)

9

From astronomy to architecture

The Project of Building is as natural to Mankind as to Birds, and was practised before the Floud. By *Josephus* we learn that Cain built the first City, *Enos,* and enclosed it with Wall and Rampires; and that the Sons of Seth, the other son of Adam, erected two Columns of Brick and Stone to preserve their Mathematical Science to Posterity, so well built that thô ye one of Brick was destroy'd by the Deluge. ye other of Stone was standing in ye time of *Josephus.*
  Wren's 'Discourse on architecture' ('Tract V'), Bolton & Hendry (1923), vol. 20, p. 140.

That Wren referred to building as a 'mathematical science' is both natural and significant. When he wrote to William Brouncker in 1663 about the entertainment of the King on a proposed visit to the Royal Society, he mentioned his 'Designes in Architecture', along with his recent work in geometry, astronomy, perspective, optics, engines and navigation (see Wren to Brouncker, 30 July 1663, Royal Society MS EL. W. 3 no. 3). We have seen that Wren's scientific interests were contained largely within the traditional domain of the mathematical sciences – geometry, astronomy, navigation, surveying, etc. – and their associated instrumentation, and architecture had always been seen as part of this domain.

Wren's colleagues at the Royal Society. no less than the earlier practitioners, were concerned with architecture. This is most obviously true of Hooke and Evelyn, but architecture was also a minor interest of Wallis, Petty, John Hoskins, Jonas Moore and John Aubrey (see Bennett, 1974, p. 307). Along with the Royal Society's work in astronomy and navigation, Sprat mentioned that: 'They have studied the promoting of Architecture in our Island' (Sprat, 1667, p. 149).

In April 1663, Wren showed the Royal Society a model of his design for the Sheldonian Theatre at Oxford

and was desired to give in writing a scheme and description of the whole frame of it, to remain as a memorial among the archives of the society.[1]

After the Fire of London, Hooke submitted to the Society, his plan for rebuilding the City[2] and Oldenburg was disappointed that Wren had not presented his (see A. R. Hall & M. B. Hall, 1965, vol. 3, pp. 230–1). Evelyn and Petty also prepared plans.[3] The Society's interests included building materials, their strengths and manufacture, and questions of structural statics.[4] The architectural taste of individual members, in so far as we know them, lent towards the ancient (classical) rather than the modern (Gothic). Evelyn (see Evelyn, 1706) and Sprat (see Sprat, 1668, pp. 242–54) were firmly of this view, and Oldenburg prefaced a volume of the *Philosophical transactions* with an argument that the so-called 'new philosophy' had ancient origins:

For this New Philosophy we were disciplined by the laudable Examples of the most Ancient Sages of the East. And we had the same or the like Guides (no less than the same Old Authority) to licence our addiction to the Mathematicks and Mechanicks, for Military, Civil, or Naval Architecture.

Philosophical Transactions 1671, **6**, 2087.

The Royal Society were anxious to emphasize their importance to society at large, without conspicuous success in the early days, and the mathematical science of architecture had a very obvious social relevance. Oldenburg thought that, if Wren's plan for rebuilding London had been submitted to the society, this

would have given the Society a name, and made it popular, and availed not a little to silence those, who aske continually, What have they done?

A. R. Hall and M. B. Hall (1965), vol. 3, p. 231.

Wren drafted a preamble (not the one adopted) to a charter incorporating the Royal Society, in which he stressed their social role:

The Way to so happy a Government, we are sensible is in no Manner more facilitated than by the promoting of useful Arts and Sciences, which, upon mature Inspection, are found to be the Basis of civil Communities, and free Governments, and which gather Multitudes, by an Orphean Charm, into Cities, and connect them in Companies.[5]

He wrote later of architecture's role in this programme:

Architecture has its political Use; publick Buildings being the Ornament of a Country; it establishes a Nation, draws People and Commerce; makes the People love their native Country, which Passion is the Original of all great Actions in a Common-wealth.

C. Wren (1750), p. 351.

Evelyn agreed on the social significance of architecture, and added aesthetic considerations to the equation:

It is from the asymmetrie of our Buildings, want of decorum and proportion in our Houses, that the irregularity of our humors and affections may be shrewdly discern'd.

Fréart (1664), dedication (to Sir John Denham).

Sprat similarly linked the need for improvements in 'the Union of our Minds, the Smoothness of our Manners, and the Beauty of our Buildings' (Sprat, 1668, p. 251).

Wren's transition to architecture was not sudden and was never complete. He was working on architectural commissions – such as the chapel at Pembroke College, Cambridge and the Sheldonian Theatre at Oxford – from the early 1660s. Indeed, the *Parentalia* list (see C. Wren, 1750, pp. 198–9) of his early work, much of which dates from the 1650s, includes 'New Designs, tending to Strength, Convenience, and Beauty in Building', along with other thoroughly Vitruvian entries, such as 'Some Inventions in Fortifications', 'Divers new Engines for raising Water', 'New offensive, and defensive Engines', 'Secure and speedier Ways of attacking Forts than by Approaches and Galleries' and 'To build in the Sea, Forts, Moles, &c'.

Wren's biographers have always seen the 1661 commission to supervise the construction of a fortified harbour at Tangier as evidence of an awakening interest in architecture. It is rather the case that the task required someone proficient in the mathematical sciences, and so reflects Wren's past experience as much as his future prospects. It is significant that Thomas Digges was called on to supervise the fortification of Dover harbour (see Johnson, 1937, p. 179) and that, when Wren declined the Tangier offer, the post passed to the surveyor and mathematical practitioner Jonas Moore.

If we turn to the late 1660s, Wren was still based in Oxford in 1668 (see Bolton & Hendry, 1923, vol. 13, pp. 46–9), although he was by then a Commissioner for rebuilding the City and was becoming increasingly involved with redesigning St. Paul's. Even in 1669, when he accepted the Surveyorship on Sir John Denham's death, Wren did not resign as Savilian Professor, but appointed Edward Bernard his deputy, until his career was more firmly decided.[6] He, in fact, resigned in 1673. Birch's *History* gives ample evidence of Wren's continued interest in the Royal Society's meetings, and he was an active president from January 1680/1 till November 1682. Hooke's *Diary* supplements this with an informal account of wide-ranging discussions among friends and in various 'clubs', one of which met regularly at Wren's home for some time during 1676. There is plenty of evidence that Wren never lost his interest in natural philosophy, or his reputation within the scientific community. After 1669, when Oldenburg had occasion to tell a foreign correspondent that Wren was too busy to be consulted on some question or other, he did not say that Wren had forsaken science for architecture, but rather that he was preoccupied with official business.[7]

Wren's move into architecture should be understood as professional rather than intellectual. Architecture had for long been accepted as part of the mathematical sciences and Wren had broad interests within this domain, even though his professional concern was astronomy. As Surveyor-General, however, his official position in architecture involved a great many administrative duties at the Board of Works and his public service left him considerably less time for his other interests.

A coincidence of circumstances helps to explain why Wren changed his profession. The social importance of the mathematical sciences had always been stressed by the English practitioners. The Royal Society, Wren among them, were now echoing this claim for their own approach to natural philosophy. Wren had only recently returned from a stimulating visit to France, and was finding new fulfilment in architectural design (see his letter to Dean Sancroft in Bolton & Hendry, 1923, vol. 13, p. 45), when the social relevance of architecture was given dramatic emphasis with the Fire of London. Hooke's background was very similar to Wren's and he too was subsequently drawn increasingly into architecture.

Attitudes among historians to Wren's visit to France in 1665–6 indicate the problems that stem from unhistorical distinctions. The visit[8] had rightly been seen as important in the development of Wren's architectural imagination, but it is wrong to say that 'the subject for study was architecture not science' (Briggs, 1953, p. 38), for he was not aware of any dichotomy of the kind.

If we look at all the available evidence, and not just the well-known letter printed in *Parentalia,* it is clear that Wren's excitement at discovering French architecture was part of his wider Vitruvian interests – in, for example, the design of French coaches (see T. Birch, 1756, vol. 2, pp. 66, 74), the 'Engines & Methods' used by the builders (see Bolton & Hendry, 1923, vol. 13, p. 17), or the construction of the quay in Paris (see Bolton & Hendry, 1923, vol. 18, p. 180). Wren's projected 'Observations on the present State of Architecture, Arts, and Manufactures in France' (see C. Wren, 1750, p. 262) was an enterprise worthy of a Fellow of the Royal Society. Virtuosi, such as Adrien Auzout, Henri Justel, Pierre Petit and Melchisédech Thévenot, played a large part in entertaining Wren in Paris. He was received as a distinguished visiting *savant,* was introduced into their circle, attended their meetings, and was shown what was considered to be of interest.[9] They helped to arrange his introduction to Bernini in August 1665.[10] Justel wrote in January 1665/6 that he was seeing Wren practically every day (see A. R. Hall & M. B. Hall, 1965, vol. 3, pp. 11–12).

The astronomer Adrien Auzout deserves a special mention. Auzout and Wren must have been known to each other as early as 1659 through Pascal's *Lettres de Dettonville* and their mutual interest in the quadrature of the cycloid (see Pascal, 1659, 'Traitte general de la roulette', p. 6). Early in 1665, Wren had studied Auzout's work on the comet of 1664–5.[11] By July, Auzout had learned of Wren's imminent visit to Paris, and wrote to Oldenburg: 'Je me reiouis de ce que nous verrons icy dans peu votre scavant M. Wren' (A. R. Hall & M. B. Hall, 1965, vol. 2, p. 428). It seems that Wren and Auzout were together fairly often, and that they discussed various subjects of mutual scientific interest.[12] We know also that Auzout had a keen interest in architecture. Lister says that Auzout had 'studied Vitruvius more than 40 years together, and very much upon the place at Rome', and, again, that 'Monsieur d'Azout was very Curious and Understanding in Architecture; for which purpose he was 17 years in Italy by times' (Lister, 1699, pp. 30, 102). After Wren had left France, Auzout was anxious to hear of his safe arrival: 'nous atendons tousiours des nouveles de larrivé de M. Wren' (A. R. Hall & M. B. Hall, 1965, vol. 3, p. 113), and Oldenburg found that Wren was 'very well satisfied wth ye civilities, he has received in France, and commends particularly Mr. Auzout' (A. R. Hall & M. B. Hall, 1965, vol. 3, p. 48).

All the evidence indicates that Wren's subsequent appointment as Surveyor-General was appropriate to a talented and practical geometer and, under the circumstances, the office was one he could reasonably accept. Nor is there much evidence that anyone found the choice strange. John Webb did protest, but he wanted the job for himself, and even his objections were based on Wren's inexperience with the administrative procedures of the Board of Works (see Bolton & Hendry, 1923, vol. 18, p. 156). There was, of course, no specific professional training for an architectural post of such standing, and Wren brought to it accomplishments expected of the traditional mathematical practitioner – geometry, draughtsmanship, a practical sense, a knowledge of surveying and mechanics, and a conviction of the social importance of the mathematical sciences.

Wren's background in the mathematical sciences, while being at the time an appropriate training for architectural appointment, was also a fundamental, if neglected, influence on the kind of architecture he produced.

In the following chapter, we will see how attitudes, characteristic of the mathematical sciences tradition in England, are reflected in some of Wren's fundamental ideas in the theory of beauty. These ideas illuminate his general approach to aesthetic problems and, in examining these more profound questions, we will rely heavily on the established conclusions of architectural criticism. For the present, however, it is interesting to find that there is one limited area where we can follow Wren's developing solution to a design problem that involved aesthetic, structural and practical considerations, and where we can recognize criteria already familiar from his early work.

One of the first problems Wren encountered as a professional architect in *c.* 1670 was one he would face over fifty times in the space of some 20 years. This was to provide an adequate and appropriate setting for worship in a London parish. Wren's City churches are solutions to this problem, variously influenced by different sets of initial conditions and restraints. Restraints involved, for example, the contribution the parish could afford to make towards costs, the size and shape of the site, and whatever remains of the former church had survived the Fire of London. The variety of these conditions means that the designs have very different degrees of significance.

In assigning dates to the designs, the building accounts[13] are only a very rough guide, but they do allow us to pick out certain lines of development in Wren's solutions and to find parallels with his attitudes to work in natural philosophy. Wren's responsibility for detail in the finished churches was probably quite limited. It is generally thought that he usually provided an overall design, without directly supervising construction. This work was more often done by assistants, most notably Hooke. (For some suggestions regarding Hooke, see Downes, 1971, pp. 151–2). We will simply look at general conceptual features of Wren's solutions, and neglect decorative elements almost entirely.

The success of any building depended, said Wren after Vitruvius, on it meeting at once the three criteria of 'Beauty, Firmness and Convenience' (C. Wren, 1750, p. 351). We have seen already that Wren held that the foundations of beauty rested in geometry,[14] though we will leave an examination of precisely what Wren meant by this dictum to the next chapter.

To some extent the geometries of beauty and firmness were linked for Wren:

Position is necessary for perfecting Beauty. There are only two beautiful
Positions of straight Lines, perpendicular and horizontal: this is from Nature,
and consequently Necessity, no other than upright being firm. Oblique
Positions are Discord to the Eye, unless answered in Pairs, as in the Sides of

an equicural Triangle: therefore Gothick Buttresses are all ill-favoured, and were avoided by the Ancients, and no Roofs almost but spherick raised to be visible, except in the Front, where the Lines answer; in spherick, in all Positions, the Ribs answer.

C. Wren (1750), p. 352.

So the pediment, to be viewed only from the front, was visibly firm and its geometry visually pleasing; and a dome generalized this success for an interior which might be viewed from any position.

These ideas, combined with Wren's basic belief in the dependence of beauty on geometry, would have led naturally to centralized designs allowing maximum symmetry. There was, however, the principle of convenience. Most of the sites Wren inherited had housed Gothic churches. In some cases he had even to use the same foundations. So he almost always began with a longitudinal space.

A second aspect of 'convenience' concerned Wren's perception of the function of a church, and how its design might be adapted to function. The problem had rarely been faced in England since the Reformation and we will be better placed to appreciate Wren's attitude after looking at the churches. But, for some preliminary guidance, his clearest statement came in a letter of advice to the Commissioners appointed under the Building Act of 1708:

The Churches . . . must be large; but still, in our reformed Religion it would seem vain to make a Parish-church larger, than that all who are present can both hear and see. The Romanists, indeed, may build larger Churches, it is enough if they hear the Murmur of the Mass, and see the Elevation of the Host, but ours are to be fitted for Auditories.

C. Wren (1750), p. 320.

A great many churches were begun in 1670[15] and reveal a variety of initial ideas.[16] The most centralized design was St Mary-at-Hill (see Figure 7), where the previous remains presented him with a rough square.[17] There are four central columns, with entablatures running from each to pilasters at the two adjacent walls, and flat ceilings over the corners. The entablatures carry barrel vaults which form a Greek cross, and a low dome over the crossing is carried on pendentives between the main arches.

The design had a great conceptual unity for Wren: he described this type of vault as 'of all others the most geometrical', being 'composed of Hemispheres, and their Sections only' (C. Wren, 1750, p. 357). If we imagine a hemisphere whose base rests on the four columns, the vertical sections through the columns will mark out the main arches, and the horizontal section above will be the base of the dome. Though he could find examples of such vaults only in Byzantine and not in ancient architecture, Wren said that: 'I question not but those at Constantinople, had it from the Greeks before them, it is so natural.'

It was, clearly, with strongly centralized designs such as this one that Wren could best illustrate his geometrical premiss. At St Benet Fink (1670, see Figures 8 and 9) he adapted the idea to accommodate an awkward site. Here the walls form a flattened decagon, the main arches spring from six columns, entablatures carry six barrel vaults, and pendentives between the arches support an elliptical dome with a lantern.

This particular solution was ingenious, but Wren did not maintain a narrow experimentation with centralized designs. Rather, he sought to accomodate their ideals to a more longitudinal plan, a move already evident at St Benet Fink. The centralized form was not versatile in dealing with seating problems – the church was, after all, an auditory – nor with the empirical restraints of most sites.

St Mary Aldermanbury (1670, see Figures 10 and 11) has a nave and aisles, separated by two rows of four detached columns. Flat ceilings cover the aisles, while columns carry entablatures supporting a barrel vault over the nave. This was a general form that Wren used on a good many

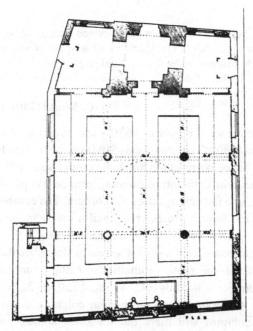

Fig. 7. St Mary-at-Hill, plan. (From Clayton, 1848.)

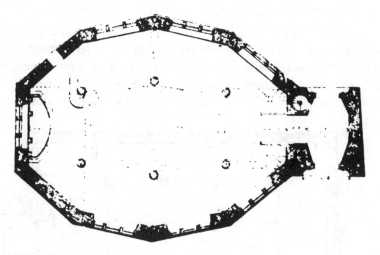

Fig. 8. St Benet Fink, plan. (From Clayton, 1848.)

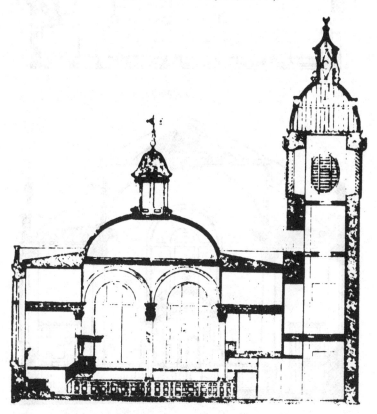

Fig. 9. St Benet Fink, longitudinal section. (From Clayton, 1848.)

occasions. Among churches begun in 1670 it appears again at St Dionis Backchurch, and in 1671 at St Magnus and at St George Botolph Lane. Here was a versatile design, which was 'convenient' in Wren's sense, but which also displayed some geometrical or 'conceptual' integration.

Yet in each case, Wren reveals the tension between beauty and

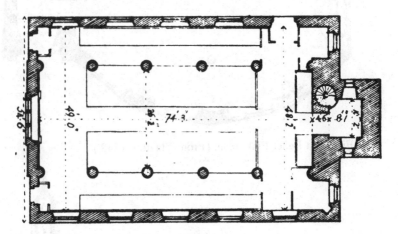

Fig. 10. St Mary Aldermanbury, plan. (From Clayton, 1848).

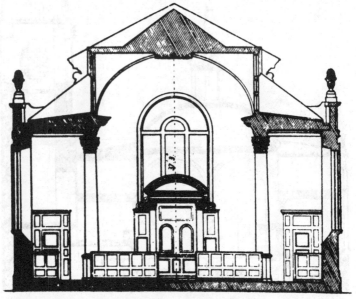

Fig. 11. St Mary Aldermanbury, transverse section.
(From Clayton, 1848.)

convenience by introducing some feature marking a north-south axis. At St Mary Aldermanbury round-headed clerestory windows are groined into the vault over the central bays; St Magnus originally had some similar features (see Summerson, 1952); at St Dionis and St George, the centralized feature is the positioning of the four columns in each case. The tension between beauty and convenience seems reflected in what will be a continued struggle between central and longitudinal forms.

So far we have seen some of Wren's ideas for small interiors. There were, of course, a good number of small and plain churches that we need not mention. But the challenge of the large longitudinal space came first with St Lawrence Jewry (1670, see Figures 12 and 13). Here Wren was designing for a wealthy parish and could have indulged some experimentation. He very carefully contrived a regular rectangular space from the irregular site, but the ceiling holds few clues to his ideas of the relations between beauty, firmness and geometry. There are explicit structural illusions: the Corinthian

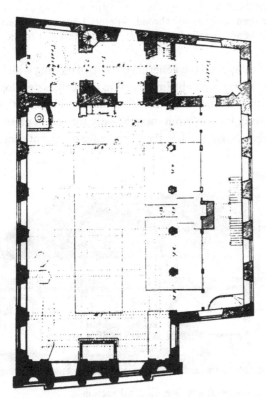

Fig. 12. St Lawrence Jewry, plan. (From Clayton, 1848.)

columns dividing off the single aisle are reflected by pilasters on the walls and, above them, moulded bands spring from the entablature like semi-arches to support the flat decorated ceiling. But, compared with his later solution to vaulting the large longitudinal space, this is a very tentative beginning.

St Mary-le-Bow (1671, see Figures 14, 15 and 16) was, however, a real point of departure. So far, Wren's basilican plans have been fairly modest – low barrel vaults over the nave, flat ceilings over the aisles. A more ambitious interior would involve a more impressive central vault and vaulted aisles. Wren had, however, already noticed a problem that he would seek to avoid. In his report on the fabric of Salisbury Cathedral (1668) he said:

Almost all the Cathedrals of the Gothick Form are weak and defective in the Poise of the Vault of the Ailes . . . True indeed, the great Load above the Walls and Vaults of the Navis, should seem to confirm the Pillars in their perpendicular Station, that there should be no need of the Butment inward; but Experience hath shewn the contrary, and there is scarce any Gothick Cathedral, that I have seen, at home or abroad, wherein I have not observed the Pillars to yield and bend inwards from the Weight of the Vault of the Aile.[18]

This particular difficulty, and the general problem of dealing with lateral thrusts generated by vaults, would be important to the further development of church designs. This was true even when structural problems of that sort were not really involved, for the 'vault' may be only formed in plaster and purely for interior effect. None-the-less the appearance was what concerned

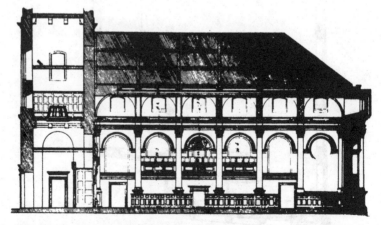

Fig. 13. St Lawrence Jewry, longitudinal section.
(From Clayton, 1848.)

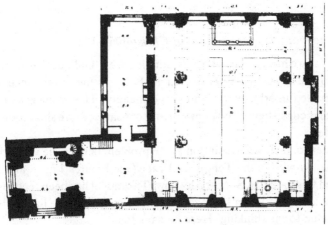

Fig. 14. St Mary-le-Bow, plan. (From Clayton, 1848.)

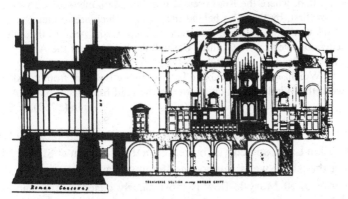

Fig. 15. St Mary-le-Bow, transverse section. (From Clayton, 1848.)

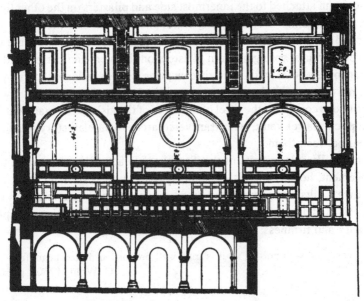

Fig. 16. St Mary-le-Bow, longitudinal section. (From Clayton, 1848.)

Wren. Just as we have seen in the most simple case – 'Oblique Positions are Discord to the Eye, unless answered in Pairs' – the appearance of a correctly poised interior was a criterion of beauty. This emphasis on the appearance is consistent with Wren's theory of beauty, as we shall see in the next chapter.

Wren knew of a classical solution to the problem of the aisle vault in the basilica then known as the Temple of Peace (see Figures 17, 18 and 19), and he used this model at St Mary-le-Bow. The model has been recognized since *Parentalia*, but Wren's own commentary on the Temple has not been used as a clue to his thinking. Here he says that

The Walls are thin, where the Roof presses not; but admirably secured where the Weight lies; first, by the Piles behind the Pillars, which are of that Thickness backward, that they are sufficient Butment to the Arch of the Ailes: (this not being observed in the Gothick Cathedrals, the Vault of the Ailes resting against the Middle of the Pillars of the Nave, bend them inwards . . .) Secondly, the Weight of the Roof above hath a mighty Butment from the slope Walls between the Windows, which answer to the Half-frontpieces of the Ailes.

C. Wren (1750), p. 362.

Wren had used half-pediments ('Half-frontpieces') in the facade of St George Botolph Lane, and curved butresses in the elevation at St Mary-le-Bow reflect the 'slope walls' supporting the nave vault.

The interior of St Mary-le-Bow is most clearly based on the Temple. There are four central pillars, each composed of a very large pier with a semi-column attached to the innermost side and pilasters on the other three sides. Longitudinal arches spring from pilasters on the piers and are continued into the aisles as barrel vaults. Into these barrel vaults are cut transverse arches, springing from pilasters on the piers and corresponding pilasters on the outer walls. The aisle vaulting is thus similar to the Temple. In the nave the semi-columns are continued up to a broken entablature with a continuous cornice at the level of the roofs of the aisles. The main vault is an elliptical barrel vault, with transverse arches and clerestory windows groined between them.

Through Wren's critique of the Temple of Peace we know that, in this arrangement, he saw the aisles supporting an impressive main vault and their own vaults being supported by piers, so as not to interfere with the nave. His model was, he thought, a successful concurrence of the principles of beauty and firmness:

as it is vast, and well poised, so it is true, well-proportioned, and beautiful.

C. Wren (1750), p. 363.

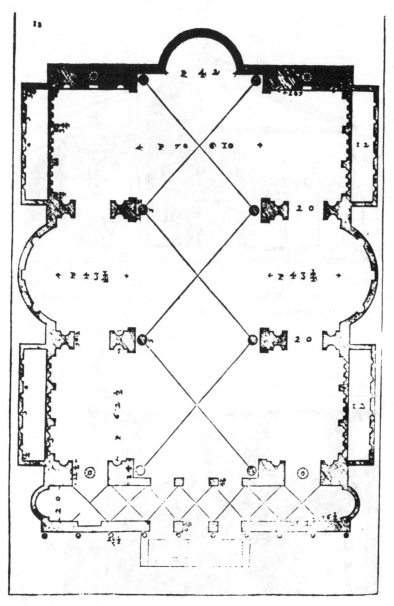

Fig. 17. The 'Temple of Peace', plan. (From A. Palladio, 1735,
Architecture, in four books, vol. 2, translated by
E. Hoppus, London.)

Fig. 18. The 'Temple of Peace', transverse section.
(From A. Palladio, 1735, *Architecture, in four books,* vol. 2,
translated by E. Hoppus, London.)

Fig. 19. The 'Temple of Peace', longitudinal section.
(From A. Palladio, 1735, *Architecture, in four books,* vol. 2,
translated by E. Hoppus, London.)

Centralized features appear also in this design, for the middle longitudinal arches are elliptical while those on either side are semi-circles.

It is interesting that St Stephen Walbrook (1672, see Figures 20 and 21) appears so early in Wren's church designs, for it was his most determined attempt to impose a centralized form on a longitudinal space. A rich parish gave him the opportunity to build a really impressive dome, almost a hemisphere, that dominates the interior, yet the experiment would not be improved on or repeated.

Wren wrote of the type of vault used at St Mary-at-Hill that 'whereas a Sphere may be cut all manner of Ways, and that still into Circles, it may be accommodated to lie upon all Positions of the Pillars'. At St Stephen Walbrook there are sixteen detached Corinthian columns, eight of which lie at the corners of a regular octagon. Wren, as we saw before, imagined this octagon the base of a hemisphere, so that the vertical sections through the columns traced out the arches and the horizontal section above was the base of the dome. The arches in the four cardinal directions open into vaults; each oblique arch onto two clerestory windows.

Tensions between centralized and longitudinal forms are marked. The columns are sited so as to give an arrangement of nave and aisles; the dome is positioned to create a traditional Latin cross. In later churches the longitudinal form became dominant as Wren developed his basilican design for large interiors. But he also tried to imbue it with the geometrical principles so clearly displayed at St Stephen.

St Bride Fleet Street (1673–4,[19] see Figures 22, 23 and 24) was a step in this process – a development of the design used at St Mary-le-Bow. At St

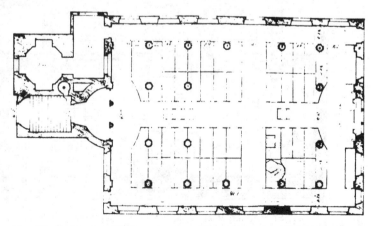

Fig. 20. St Stephen Walbrook, plan. (From Clayton, 1848.)

Bride there are five bays north and south, marked out by coupled Doric columns, which rise to a broken entablature, supporting both the longitudinal arches and the transverse arches of the aisles. At the keys to the longitudinal arches, a cornice runs the length of the church and corbels above the columns support transverse arches spanning the nave. The main semicircular vault is groined over clerestory windows between the arches. The groined ceiling over the aisles has elliptical transverse arches.

The coupled columns replace the semi-column and pier composition at St Mary-le-Bow, but reveal a similar concern – to provide lateral width to

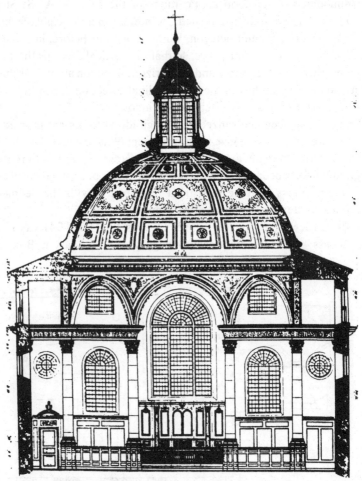

Fig. 21. St Stephen Walbrook, transverse section.
(From Clayton, 1848.)

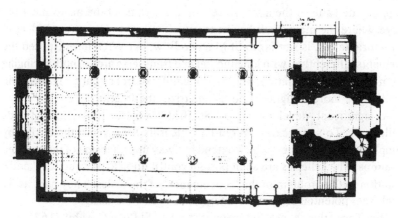

Fig. 22. St Bride Fleet Street, plan. (From Clayton, 1848.)

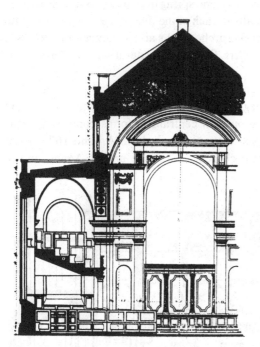

Fig. 23. St Bride Fleet Street, transverse section.
(From Clayton, 1848.)

support the vault of the aisle. Here again, the roofs of the aisles meet the nave where the main vault begins to give it support. The general idea, in structural terms, is similar to St Mary-le-Bow, but Wren has disposed the elements differently – with longitudinal and transverse arches both springing from a broken entablature, and the longitudinal arches carrying the main vault – the more to lighten and 'geometrize' the interior.

A significant development at St Bride was the provision of side galleries.[20] To some extent Wren has integrated them into the design by supporting them on piers which extend halfway up either side of the coupled columns. The galleries are an important contribution to the 'auditory' and a solution to the problem of the 'convenience' of the aisles. They had, as we shall see, potential for further integration within the design.

Wren used this general design once more, at St Peter Cornhill (1675–7,[21] see Figures 25, 26 and 27). Large piers again separate narrow aisles from the nave. A pilaster on the inside of each pier rises to a Corinthian capital, a broken entablature, and a cornice which ranges with the keys of the longitudinal arches. These arches again spring from pilasters attached to the piers and open into barrel vaults in each of the five bays on either side. Into these vaults are cut the transverse arches of the aisles; they rest on corbels at the walls and pilasters on the piers. Once again, lateral width in the pillars is coupled with vaulted aisles.

In concentrating on Wren's development of a solution to the large longitudinal space, we will neglect to deal with the smaller churches, save to say that, here, experimentation with centralized forms continued, along with compromise adaptations to longitudinal plans. Around 1676–7, for example, Wren designed two churches, St Anne and St Agnes, and St

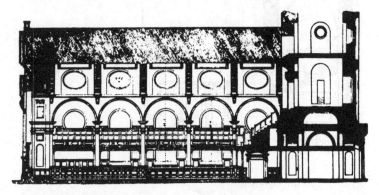

Fig. 24. St Bride Fleet Street, longitudinal section.
(From Clayton, 1848.)

Martin Ludgate, each with four columns and barrel vaults intersecting at the centre. He also designed the wonderfully ambiguous St James Garlickhythe[22] (see Figure 28), where the idea is adapted to a longitudinal premiss.

In the first decade of Wren's church designs, we see two types of 'basilican' solutions. Either a vaulted nave is separated from unvaulted

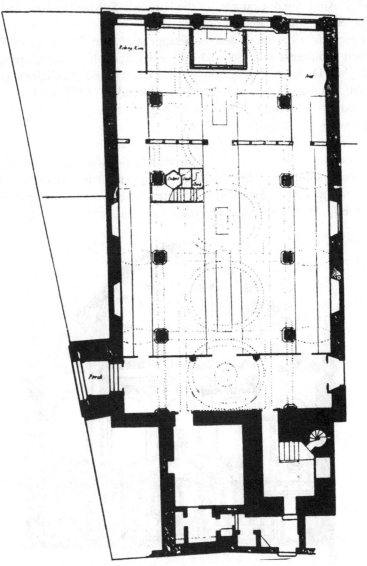

Fig. 25. St Peter Cornhill, plan. (From Clayton, 1848.)

aisles by columns, or a vaulted nave is separated from vaulted aisles by some combination of elements whose functions – at least conceptually – is to butress the aisle vault. This much is clear from Wren's own critique of the Temple of Peace. The main vault begins at the level of the aisle roofs. Examples of the former design are St Dionis, St Mary Aldermanbury, St George, St Magnus, St Michael Bassishaw, Christ Church Newgate Street and St Anne Soho; examples of the latter are St Mary-le-Bow, St Bride Fleet Street, and St Peter Cornhill.

From 1680 on, these designs disappear, and in five churches Wren adopts a different solution – which we can again link with his recorded thoughts. These are St Augustin Watling Street (1680), St Clement Danes (1680, see Figures 29 and 30), St James Piccadilly (1682, see Figures 31, 32 and 33), St Andrew Holborn (1684–5, see Figures 34 and 35), and St Andrew-by-the-Wardrobe (1685).

Formerly, in the larger interiors, the main vault had been carried by the longitudinal arches at the level of the aisle roofs. Now, vaults of both nave and aisles spring from the same place – a broken entablature carried by the columns, and the longitudinal arches are now groined into the main vault. Formerly, Wren had imagined the aisle vaulting pressing on the nave and

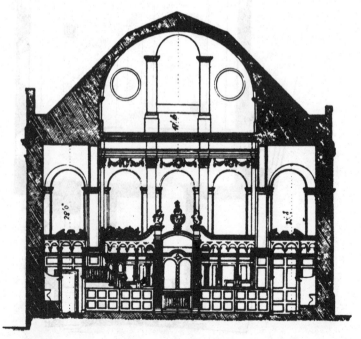

Fig. 26. St Peter Cornhill, transverse section. (From Clayton, 1848.)

had introduced compensating buttressing. Now the two vaults are poised against each other and Wren rests them both on single columns. Formerly, we saw Wren struggling with compromises between central and longitudinal forms in seeking a satisfactory solution for the large rectangular space. Now he has a design that exhibits the geometrical basis of beauty and firmness within the longitudinal form – a satisfactory 'conceptual' solution – that he uses whenever it is appropriate.

Wren was able also to integrate the side galleries into the solution. Generally, they are carried on piers with the columns rising from the gallery fronts. The windows are also accommodated to the vaulting and the galleries. Formerly, there was a lower storey and a clerestory. Now there are large round-headed windows above the galleries and smaller windows beneath.

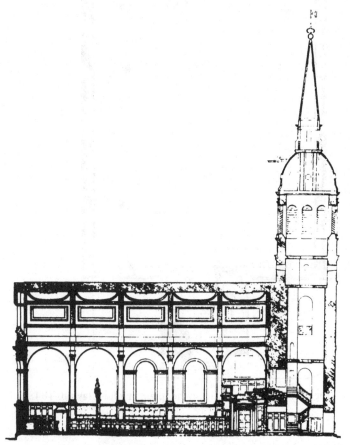

Fig. 27. St Peter Cornhill, longitudinal section. (From Clayton, 1848.)

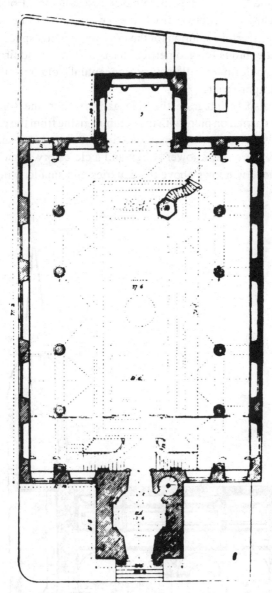

Fig. 28. St James Garlickhythe, plan. (From Clayton, 1848.)

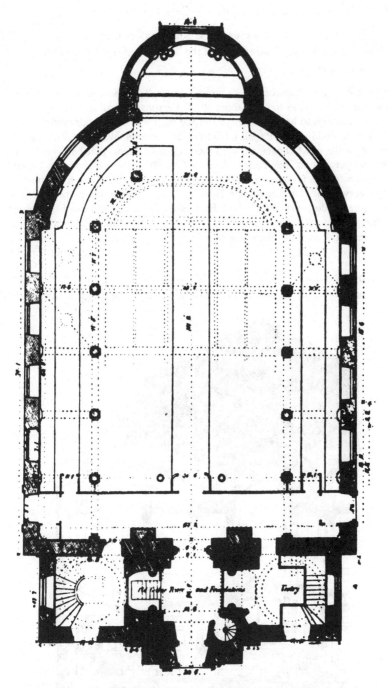

Fig. 29. St Clement Danes, plan. (From Clayton, 1848.)

In Wren's letter of advice to the Commissioners, appointed under the Building Act of 1708, he stresses that the whole congregation should be able to see and hear:

I can hardly think it practicable to make a single Room so capacious, with Pews and Galleries, as to hold above 2000 Persons, and all to hear the Service, and both to hear distinctly, and see the Preacher. I endeavoured to effect this in building the Parish Church of St. James's, Westminster, which, I presume, is the most capacious, with these Qualifications, that hath yet been built; and yet at a solemn Time, when the Church was much crowded, I could not discern from a Gallery that 2000 were present. In this Church I mention, though very broad, and the middle Nave arched up, yet as there are no Walls of a second Order, nor Lantern, nor Buttresses, but the whole Roof rests upon the Pillars, as do also the Galleries; I think it may be found beautiful and convenient, and as such, the cheapest of any Form I could invent.

C. Wren (1750), p. 320.

Wren's comments on St James, Piccadilly, are very revealing. There is an impressive vault; the church is 'very broad, and the middle Nave arched

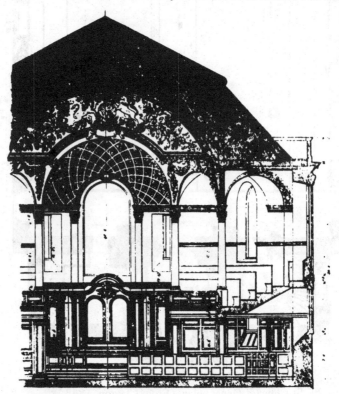

Fig. 30. St Clement Danes, transverse section. (From Clayton, 1848.)

up'. He has achieved this with 'no Walls of a second Order, nor Lantern, nor Buttresses', which is to say that the main vault is carried directly by the columns with the longitudinal arches groined into it, instead of being carried above the longitudinal arches and pierced by a clerestory. The poise of the whole vault is demonstrated by the single columns with the galleries integrated into the structure: 'the whole Roof rests upon the Pillars, as do also the Galleries'.

Wren is pointing precisely to the change he made between the earlier large basilican designs, and those begun from 1680 onwards. We saw him searching for an integrated and conceptually satisfactory solution to the

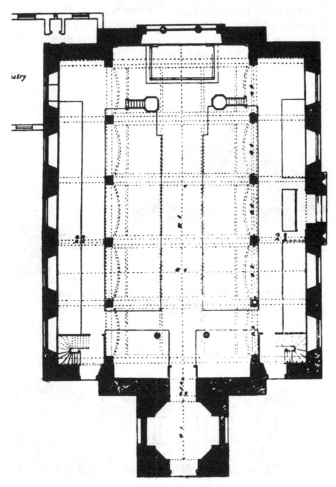

Fig. 31. St James Piccadilly, plan. (From Clayton, 1848.)

longitudinal space. This he has now found: 'I think it may be found beautiful and convenient, and as such, the cheapest of any Form I could invent.' We have treated Wren's practice as something close to invention, aimed at a kind of intellectual economy. St James was not a plain or cheap church; rather, the 'form' was neat, integrated and economical.

One of the most pervasive guiding principles of Wren's mathematics and natural philosophy was intellectual economy or neatness. This he expressed most clearly in his letter on the genesis of his double-writing instrument:

the misapprehending World measures the Excellence of things by their Rarity, or Difficulty of Framing, not by the Concinnity and apt Disposal of Parts to attain their End by a right Line as it were & the Simplest way. Any New Invention in Mechanicks perform'd by an Operose way of divers unessential though well compacted Parts, shall be admired together with the Artist, meerly for the Variety of the Motions and the Difficulty of Performance; comes a more judicious Hand and with a far smaller number of Peeces, & those perhaps of

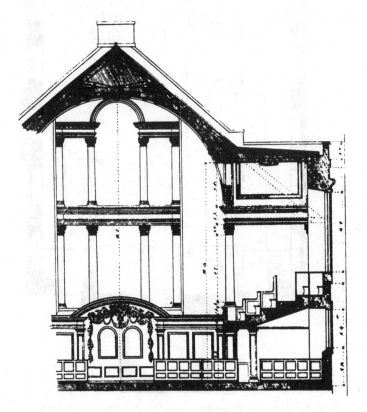

Fig. 32. St James Piccadilly, transverse section. (From Clayton, 1848.)

more trivial Materials, but compos'd with more Brain & less ostentation, frames the same thing in a little Volume, & such a one I shall call a Master.

Bennett (1973), p. 144.

Similar ideas appear elsewhere in his architecture, for in writing to Dean Sancroft about a possible design for the new cathedral, Wren said that even if they could not reproduce the scale of Old St Paul's, they might at least build 'some neate fabrick, wch shall recompence in Art and beauty what it wants in bulke' (Bolton & Hendry, 1923, vol. 13, pp. 45–6).

By treating the churches as solutions to a design problem, and by arranging these solutions in time, we have had before us material of a kind familiar in the history of science. The approach has brought out important points that have not been noticed before; and we have been able to recognize

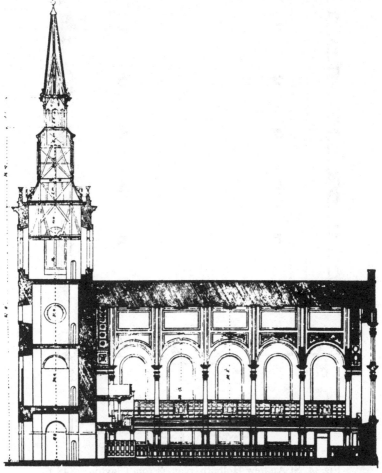

Fig. 33. St James Piccadilly, longitudinal section. (From Clayton, 1848.)

at work, criteria already encountered in Wren's natural philosophy. But this kind of technique does not take us much further. What is more important is to draw on the more broadly-based conclusions of architectural criticism and to show that the underlying philosophy was similar to that revealed in Wren's science: the philosophy of the mathematical sciences tradition in England.

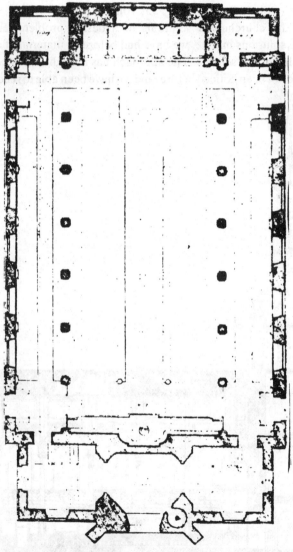

Fig.34. St Andrew Holborn, plan. (From Clayton, 1848.)

The churches are, however, useful in one other way. We have little direct record of Wren's attitudes to questions of faith and religion. He did show a rather cavalier confidence in the rational explanation of miracles[23] but, in general, there is insufficient evidence to form a very clear impression. Yet the churches may be as eloquent a statement as any: the terms in which we have seen Wren approach church design and the requirements he placed on a satisfactory solution.

In 1668 Thomas Sprat published a book addressed to Wren, in which he defended England against the criticism of a French visitor (see Sprat, 1668). He recalled how he and Wren 'have sometimes debated together, what place and time of all the past, or present, we would have chosen to live in, if our fates had bin at our own disposal', and how 'we both agreed, that Rome, in the Reign of Augustus, was to be preferr'd before all others'. The values and interests of the court, they believed, had at that time created an environment 'nothing pleasanter to a Philosophical mind'.

Since the Restoration, however, Sprat continued, the Augustinian age had been reborn in England. A just and open society, an enlightened and moderate government, a quickening of natural philosophy were all allied to 'the Profession of such a Religion, and the Discipline of such a Church, which an impartial Philosopher would chuse'. The sentiment is perfectly reflected in Wren's churches. In making the Temple of Peace a point of departure, Wren had chosen, in his own words, 'a Hall of Justice, and for that Reason it was made very lightsome whereas the consecrated Temples were generally very obscure' (C. Wren, 1750, p. 362).

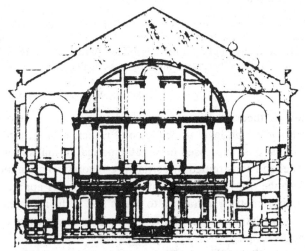

Fig. 35. St Andrew Holborn, transverse section. (From Clayton, 1848.)

10

The natural causes of beauty

We began in Chapter 1 with two general problems: Wren's switch of interests from astronomy to architecture; and the relation between his architecture and his early work in natural philosophy. The mathematical sciences tradition in England has helped to explain the domain he accepted for his work and to show that there was no fundamental shift in intellectual terms, even though the change had far-reaching professional implications for Wren. The same tradition is relevant to understanding the character of Wren's work in both professions.

We have seen clearly how fundamental was the role which mathematics, as it was understood within the mathematical sciences, played in Wren's natural philosophy. Its influence can be traced in the forms of abstract regulative criteria, direct mathematical formulation, and practical maths-based technology. Echoing the creeds of early practitioners like Recorde or Digges, Wren wrote that:

Mathematical Demonstrations being built upon the impregnable Foundations of Geometry and Arithmetick, are the only Truths, that can sink into the Mind of Man, void of all Uncertainty; and all other Discourses participate more or less oƒ Truth, according as their Subjects are more or less capable of Mathematical Demonstration.

C. Wren (1750), pp. 200–1.

Yet the kinds of theories Wren formulated were not mathematical derivations from some high-level principles but, rather, were generalized accounts of observations, expressed in a mathematical formalism or as a geometrical model and constrained by certain pre-theoretical assumptions of neatness, elegance or symmetry. Examples would be his approach to comets, to pendulum motion, to Saturn, to optics and to the mechanics of sailing, in so far as we have any record and – in the best-documented case – to elastic impact.

We saw in Chapter 7 that Wren's 'Theory of Motion' consisted of rules, expressed in a formalism derived from the balance that synthesized his experimental results, and that he presented these rules with no attempt at an explanation or derivation. This deficiency, while a problem for some of his critics, was stoutly defended by Wren who held that his rules were justified rather by the phenomena.

Huygens, Leibniz and William Neile all voiced a common criticism: Wren's rules were true and accurate, but were only empirical rules with no satisfactory general explanation.[1] Wren's answer was that the rules were justified by experiment, and that a supposed demonstration would assume propositions more unknown than the rules themselves (see A. R. Hall & M. B. Hall, 1965, vol. 5, pp. 347, 375). Neile, for example, reported:

I wish Dr. Wren would explain his principles a litle more fully but his is against finding a reason for the experiments of motion (for ought I see) and says that the appearances carrie reason enough in themselves as being the law of nature.

A. R. Hall & M. B. Hall (1965), vol. 5, p. 263.

In general, we can say that Wren used mathematics and related criteria of neatness and elegance as regulative principles in theorizing, but that he did not underpin these working principles or guides with any developed metaphysics. The theory was justified from below rather than from above.

Wren often envoked these unwritten regulative guides by using the term 'natural'. He tells us, for example, that a solution to the Saturn problem should be 'agreeable to the uniform and beautiful harmony of natural motions' (Van Helden, 1968, p. 220), and that his model was 'so simple and natural, depending solely on the rotation or inclination of the body' (Van Helden, 1968, p. 222). He later adopted Huygens's alternative on the grounds of 'neatnesse . . . & Naturall Simplicity' (Royal Society MS EL. W. 3 no. 2). In his theory of motion, he calls the velocities in the fundamental case of a 'balanced' collision, 'proper and most truly natural' (A. R. Hall & M. B. Hall, 1965, vol. 5, p. 320). In architecture we have seen already that Wren felt that the Greeks must have invented the Byzantine idea of carrying a dome on pendentives, 'it is so natural' (C. Wren, 1750, p. 357) (this vault was 'of all others the most geometrical'); and the central problem of this chapter will, of course, be Wren's concept of 'natural beauty'.

That Wren did not seek to justify his belief in the importance of mathematics through any kind of high-level universal or metaphysical principles is, we have seen in Chapter 2, typical of the mathematical sciences in England – 'Vitruvian' in character rather than 'Platonic'. Mathematics was more a tool to be applied than a privileged source of special enlightenment. Wren's attitude to the ontological status of mathematics is clear from his criticism of astrologers for 'ceremoniously numbering the critical Days, not considering that neither Time or Number hath any reality *extra intellectum humanum*' (C. Wren, 1750, p. 202).

Wren's theory of beauty – a proposed basis for his practice in architecture – is thoroughly consistent with his approach to the other mathematical sciences. The key paragraph from his first architectural tract was:

> There are natural Causes of Beauty. Beauty is a Harmony of Objects, begetting Pleasure by the Eye. There are two Causes of Beauty, natural and customary. Natural is from Geometry, consisting in Uniformity (that is Equality) and Proportion. Customary Beauty is begotten by the Use of our Senses to those Objects which are usually pleasing to us for other Causes, as Familiarity or particular Inclination breeds a Love to Things not in themselves lovely. Here lies the great Occasion of Errors; here is tried the Architect's Judgment: but always the true Test is natural or geometrical Beauty.
>
> C. Wren (1750), p. 351.

In Chapter 1, we saw the problems that stem from the standard interpretations of natural and customary beauty, but we are now better placed to suggest alternatives.

Wren sought architectural principles that would withstand changing times and fashions:

> Architecture aims at Eternity; and therefore the only Thing uncapable of Modes and Fashions in its Principals, the Orders.[2]

If buildings are to last for centuries, their appeal must depend, not on current fashion, but on some more fundamental principles of beauty. Customary beauty has generally been understood as a sanction for imaginative expression and as providing the theoretical basis for Wren's affinity with aspects of the Baroque. In fact, it is clear that he introduces the concept as a warning, which must be understood in terms similar to Bacon's 'Idols' – those influences which confuse and mislead the mind in the search for truth, and which must be systematically removed before the mind is free to discover the true system of nature.

Wren began with an empirical premiss – 'Beauty is a Harmony of Objects, begetting Pleasure by the Eye' – a premiss that defines beauty in terms of the visual experience of the observer. But 'customary' causes of a pleasurable experience – 'as Familiarity or particular Inclination' – must not guide the architect who aims at eternity.

Wren himself used his notion of customary beauty in comments on certain buildings. Familiarity, he thought, was the primary influence on judgments of Gothic buildings:

I have found no little difficulty to bring Persons, of otherwise a good Genius, to think anything in Architecture could be better then what they had heard commended by others, and what they had view'd themselves. Many good Gothick forms of Cathedrals were to be seen in our Country, and many had been seen abroad, which they liked the better for being not much differing from Ours in England.

Bolton & Hendry (1923), vol. 19, p. 140.

'Particular Inclination', on the other hand, included novelty, 'in which', Wren said, 'Fancy blinds the Judgment' (C. Wren, 1750, p. 352). In his letter from France, he wrote:

the Women, as they make here the Language and Fashions, and meddle with Politicks and Philosophy, so they sway also in Architecture . . . but Building certainly ought to have the Attribute of eternal, and therefore the only Thing uncapable of new Fashions.

C. Wren (1750), p. 261.

Yet Wren was confident that there were 'natural Causes of Beauty', that, cleansed of disturbing influences, men would make consistent aesthetic judgments. He once assured Roger North 'that there was that distinction in Nature of graceful and ugly; and that it must be so to all creatures that had vision' (Colvin, 1951, p. 259), and he wrote in 'Tract I': 'Geometrical Figures are naturally more beautiful than other irregular; in this all consent as to a Law of Nature' (C. Wren, 1750, p. 351). The architect must therefore aim to exploit these natural causes, for the customary would not last.

It is no surprise that for Wren 'Natural is from Geometry'. Mathematics was, after all, the basis of all certainty, 'and all other Discourses participate more or less of Truth, according as their Subjects are more or less capable of Mathematical Demonstration'. Architecture was a mathematical science.

But, like customary beauty, natural beauty is subject to Wren's empirical premiss. Strictly speaking, he does not refer to 'kinds' of beauty but to 'causes' of beauty – some causes are transient, others more lasting. It is not that geometry has some universal meaning of significance in itself; rather, Wren is confident that man's natural judgment, once rid of the disturbing effects of familiarity and fashion, will prefer regular geometrical forms and visually balanced proportions: 'always the true Test is natural or geometrical Beauty'. He used something like this concept when writing to Dean Sancroft in May 1666 concerning the restoration of the old cathedral of St Paul's:

I hope you will goe to the charges of trew latine . . . Take one consideration; How that gives you a well-projected Designe, opens his heart to you, and tells you all at first.   Bolton & Hendry (1923), vol. 13, p. 44.

Wren's natural beauty has been identified with the concept of Renaissance theorists such as Vignola, Alberti or Palladio, for whom correct geometrical proportions could be objectively defined in terms of harmonic relations of a cosmic order. This is not so: Wren did not believe in a definitive geometrical grammar informing all architectural design. Beauty did not subsist in some objective way in terms of the geometrical relations exhibited by the design. Just as Wren had condemned astrologers for 'ceremoniously numbering the critical Days', he criticized architects who tried to reduce the proportions of the orders to rules 'too strict and pedantick'.[3] His natural beauty had a much weaker basis: his confidence in man's unprejudiced judgment. In epistemological terms this was, of course, a weakness – similar to the one William Neile pointed to in connection with Wren's rules of motion, which were also mathematically expressed, but not derived:

for Dr Wren I think he assumes his axiome a great deale sooner then he need to doe . . . to conclude that the aparence is the reality and that the aparence must not be denied to be really true under pretence that it is an axiome meethinks is not very philosophicall.

A. R. Hall & M. B. Hall (1965), vol. 5, p. 363.

In architecture too Wren had no developed metaphysics underpinning his confidence in the importance of geometry, a deficiency wholly typical of the mathematical sciences. It is significant that Wren had no Platonic metaphysics in his theory of geometrical beauty, the very place where we might have expected to find one. He even demurred at Vitruvius's comparison between the proportions of the Doric column and those of a man, which, although significant, was hardly very extravagant. Wren felt rather that columns imitated trees: 'This I think the more natural Comparison, than that to the Body of a Man, in which there is little Resemblance of a cylindrical Body.' (C. Wren, 1750, p. 353).

Claude Perrault's *A Treatise on the five orders of columns in architecture*[4] has often been cited as an important source of Wren's ideas. The link with Wren is drawn from the fact that Perrault, who is arguing for a liberal attitude towards the orders, distinguishes two sorts of beauty – positive and arbitrary, the latter being based on custom. He introduces the distinction in order to disregard the strict rules of proportion by labelling them 'arbitrary' and holding that the 'positive' aspects of the beauty of ancient buildings are not their proportions. Wren, on the other hand, introduces his distinction so that he might hope to find some governing principles in architecture, though these were not strict rules of proportion.[5] Perrault's positive beauty is based on common sense; there is no mention of geometry so central to Wren's thesis. Perrault sees the true architect as one skilled in arbitrary beauty, in questions of taste and fashion, while for Wren 'the true Test is natural or geometrical Beauty'. Perrault's *Ordonnance* first appeared in a French edition of 1683, and we know only that Wren possessed the English translation of 1708 (see Bennett, 1974, p. 475). Tract I may be later than this, but Wren was using the term 'natural' in his natural philosophy in the 1650s, and writes of 'natural beauty' in his report on Salisbury Cathedral, written in 1668 (see Bolton & Hendry, 1923, vol. 11, p. 21).

Roland Fréart's *Parallel of the ancient architecture with the modern*[6] is a much more reactionary work, expressing the traditional attitude to classicism, yet the language has much in common with Wren. Fréart stresses the central importance of geometry, and he speaks of

the Symmetry and Oeconomy of the whole . . . producing as 'twere a visible Harmony and Consent, which those Eyes that are clear'd and enlightened by the real Intelligence of Art, contemplate and behold with excess of delectation.
Fréart (1664), preface, p. 3.

He refers to the 'blind Respect and Reverence' which comes from custom (Fréart, 1664, preface, p. 2), and says that the ancient Greeks, in their architecture 'discern'd those things as 'twere naturally, which we discover with so much pain, after a long and laborious indagation' (Fréart, 1664, preface, p. 4). We know that Evelyn presented Wren with a copy of his English translation of Fréart in 1665 (see Bennett, 1974, p. 442), and in fact Wren's copy of this edition is still extant. He also had the second edition of 1707 (see Bennett, 1974, p. 466).

The source of Wren's language is a secondary question, for the source of his ideas is clear. Wren's thesis, in substance, was very different from that of Fréart, and it was the traditional attitudes associated with the mathematical sciences that lay at the basis of the theory of beauty he proposed to apply to architecture.

Moving from theory to practice, we find that the same ideas will hold good. Wren's 'natural beauty' would have provided the theoretical ground for the kind of architecture his commentators have recognized, and which they have seen as, in some way, a compromise between Renaissance and Baroque. Geometry was the basis of natural beauty, so that Wren employed the classical idiom with its affinity to, and indeed its theoretical foundation in, geometry. But, lacking a metaphysics, he would not adhere to a strict set of rules, justified by and derived from some universal principles. What we would expect is just what Summerson has identified as a 'loose unconventional fashion' (see above, p. 3), which was far from pure classicism but not in sympathy with the expressive freedom of the Baroque.

The tension with which we began in Chapter 1, and which characterizes the critical accounts of Wren's architecture, is now readily understood, but it stems from the critical apparatus itself. With this perspective, Wren seems to be 'inconsistent', to 'falter' between two alternatives, or to be 'forced' towards one against his instincts.

What now appears is a remarkable consistency in the corpus of his work. The mathematical sciences was a subject area with an established tradition in England and characteristic attitudes. Wren's education and early work drew him into this tradition and it informed his work within the established sciences such as astronomy and navigation, as well as shaping his approach to the wider field of experimental natural philosophy. It became also an appropriate foundation on which to build his new career as a professional architect.

Notes

Chapter 1

1 C. Wren, 1750, pp. 351–3. A study of the evolution of the *Parentalia* and its use as a source is in Bennett, 1973; and the particular case of the Tracts on architecture is considered in Bennett, 1972, pp. 6–8.

Chapter 2

1 On the mathematical tradition in England in the period 1550–1650, and its origins, see especially E. G. R. Taylor, 1954; Johnson, 1937; Waters, 1958; C. Hill, 1965, ch. 2.
2 Recorde, 1551, preface. Wren's copy of *The pathway* is in the Bodleian Library, shelfmark Savile K.6(5).
3 Johnson, 1937, p. 132. Wren's copy of Recorde's *Castle* is in the Bodleian Library, shelfmark Savile K.5(3).
4 Wren's copy is in the Bodleian Library, shelfmark Savile M.19(4). He also had an issue of L. Digges's *Prognostication of right good effect,* Savile K.6(4).
5 Wren's copies of Fosters's *Elliptical, or azimuthal horologiography* and his *Miscellanies* are in the Bodleian Library, shelfmarks Savile I.2 and Savile N.14.
6 Yates, 1969, chs. 1–3; note especially Wilkins in *Mathematicall magick.*

Chapter 3

1 See C. Wren, 1750, p. 185; there is an English translation in Milman, 1908, pp. 19–20.
2 Shelfmarks 0.2.26 Art. Seld. and T.11.20.Th; see Colie, 1960.
3 Perhaps the most striking example is the similar arguments of Wren and Dean Wren on the interpretation of Christ spending three days and three nights in the tomb, see C. Wren, 1750, pp. 202–3, and the final page of Browne's *Pseudodoxia epidemica,* Bodleian Library, shelfmark 0.2.26 Art. Seld.
4 Browne, *Pseudodoxia epidemica,* Bodleian Library, shelfmark 0.2.26 Art. Seld., p. 18; note also *ibid.,* pp. 292, 294, 366. In his copy of Bacon's

Sylva sylvarum, the Dean argues against Galileo's tidal 'proof' of the Earth's motion, see his note opposite p. 235.

5 Holder, 1694, p. 14. Wren maintained contact with Holder long after this early period. For example, we know from Hooke's Diary that Holder was a member of the 'New Philosophicall Clubb' that began to meet at Wren's home on 1 January 1675/6.

6 Clark, 1898, vol. 1, p. 405; Clark, 1898, vol. 2, p. 285. For later links, see H. W. Robinson & W. Adams, 1935, pp. 104, 386, 428; Holder, 1678, p. 5.

7 Munk, 1878, vol. 1, pp. 252–5; Keevil, 1952; Goodall, 1684; Wallis, 1685, p. 301; Oughtred, 1652, preface; Moore, 1681*a,* p.119; Moore, 1681*b,* preface.

8 McKie, 1960, p. 12. Scarburgh's name was omitted in Wallis's later account, see Purver, 1967, p. 166.

9 See Clark, 1898, vol. 2, p. 285; see also Pope, 1697, pp. 18–19.

10 While Dean Wren was at Windsor, Charles Lewis (elder brother of Prince Rupert) spent a few weeks at his home. Wilkins was Charles's private chaplain at the time, see C. Wren, 1750, p. 183.

11 C. Wren, 1750, p. 183. C. Wren Jnr dated this letter to 1652, see British Library MS Add. 25 071, fo. 36.

12 C. Wren, 1750, pp. 198–9. As a rough check on the reliability of this list, whose source is not clear, we can note that, for over two-thirds of the entries, there is independent evidence of Wren's interest, and those we can date can be traced to the period before his Gresham appointment.

13 G. H. Turnbull, 1952, p. 111. Two early publications are Wren's poem on the revival of an executed woman, attributed to Petty, Willis & Bathurst, – Watkins, 1651; and his letter on the Wadham beehives in Hartlib, 1655.

14 Fuerst, 1956, note 802; note also Clark, 1898, vol. 2, pp. 113–14.

15 See Gardiner, 1889, p. 178. Even in 1666 Wren addressed a letter from Wadham, see Bolton & Hendry, 1923, vol. 13, p. 44.

16 G. H. Turnbull, 1952, p. 116. What Wilkins may have achieved is not known, but in 1659 Thomas Barlow wrote to Robert Boyle: 'Dr. Wilkins gone (*cum pannie*) to Cambridge, and left his great telescope to the library'. T. Birch, 1772, vol. 6, p. 301.

17 E. S. de Beer, 1955, vol. 3, p. 110. 'Mathematical magic', or the use of mechanics to produce spectacular effects or impressive feats of technology, had been recognized as part of the mathematical sciences, not only by Dee, but also much earlier by Recorde, see Recorde, 1551, preface.

18 Detailed accounts of Wren's relations with his associates in natural philosophy, with references, are given in Bennett, 1974, ch. 2.

19 T. Birch, 1772, vol. 2, pp. 88–9; *Philosophical Transactions,* 1665–6, **1,** 129; A. R. Hall & M. B. Hall, 1965, vol. 2, p. 616; Bennett, 1973, p. 147; G. H. Turnbull, 1952, p. 112.

20 For only some of the evidence of their working relations, see Royal Society MS EL.3 no. 3; H. W. Robinson & W. Adams, 1935, pp. 333, 344; T. Birch, 1772, vol. 3, p. 257.

21 For Wren's work on Saturn, see Van Helden, 1968; Ronan & Hartley, 1960; Armitage, 1960.

22 For Neile, see Ronan & Hartley, 1960. Wren maintained contact with him well into the 1660s (see note 18).

23 For Wren's contacts with Neile's mathematics, see *Philosophical Transactions*, 1673, **8**,6150; note also A. R. Hall & M. B. Hall, 1965, vol. 4, p. 55.

24 See S. Ward, 1653, preface; S. Ward. 1656, postscript to dedication.

25 We have already mentioned his visit to Ashmole with Wilkins in 1652; note also Wren in C. Wren, 1750, p. 215 and that Dean Wren was in London in 1652, see his note in his copy of Bacon's *Sylva sylvarum* at the Bodleian Library, T.11.20.Th., p. 162. Wren called on Hartlib in March 1653, and again in September 1655, see G. H. Turnbull, 1952, pp. 110, 114.

26 See Huygens, 1888, vol. 1, p. 401; Huygens, 1888, vol. 2, p. 306; Huygens, 1888, vol. 4, pp. 169, 255; C. Wren, 1750, p. 242.

27 See Wren to Neile, 1 October 1661, Royal Society MS EL.W.3 no. 2, printed in T. Birch, 1756, vol. 1, pp. 47–9; Huygens, 1888, vol. 3, pp. 416–18; Ronan & Hartley, 1960, pp. 162–4.

28 See Van Helden, 1968, pp. 223–4; Armitage, 1960, p. 167; note also Robertson, 1931, p. 168.

29 See Copeman, 1960, pp. 71–2; A. Hill, 1683, vol. 1; Stimson, 1931, p. 551. Wren was succeeded by Wilkins's half-brother, Walter Pope, who himself wrote that Wilkins and Goddard used their positions 'to advance the interest of Learning', Pope, 1697, p. 46.

30 See, for example, G. H. Turnbull, 1952, p. 112; Crossley, 1847, p. 127. We know Wren was resident at All Souls in 1659, when the lectures at Gresham were suspended, see C. Wren, 1750, pp. 254–5; McKie, 1960, p. 19; note also Frank, 1973, p. 260.

31 See Monconys, 1665, vol. 2, pp. 74–5. For grinding lenses, Wren and Neile employed Richard Reeves, who had also worked for Pell, see E. G. R. Taylor, 1954, pp. 223–4.

32 Wallis, 1685, p. 293; .see also Wallis, 1659, p. 91; *Philosophical Transactions*, 1673, **8**, 6146–9.

33 See J. Ward, 1740, p. 8. For the duties of the Savilian Professor of Astronomy, see Allen 1949, p. 226; S. Ward, 1654, p. 30.

34 See Crossley, 1847, pp. 126–7; T. Birch, 1772, vol. 4, p. 118; T. Birch, 1756, vol. 1, p. 505. The reference in *Parentalia* to 'Lecturae anglicae & latinae, de luce & refractione' may be relevant as the Gresham Professors were required to deliver their lectures in both English and Latin, C. Wren, 1750, p. 241.

35 Wren to Neile, 1 October 1661, Royal Society MS EL.W.3 no. 2; see also J. Ward, 1740, p. 97.

36 Crossley, 1847, p. 215. Neile did provide a telescope for Whitehall, see Huygens, 1888, vol. 15, pp. 70–1; Monconys, 1665, vol. 2, p. 77.

37 See J. Ward, 1740, p. 97; Clark, 1891, vol. 1, p. 380.

Chapter 4

1 See Johnson, 1937, pp. 126–8. Wren's copy of Recorde's *Castle* is in the Bodleian Library, shelfmark Savile K.5(3).

2 Van Helden, 1968, p. 220. Wren refers to the published sketches of Galileo, Fontana, Gassendi, Riccioli and Hevelius, and had some relevant works in his library, see Bennett, 1974, p. 215.

3 Van Helden, 1968, pp. 221, 225; C. Wren, 1750, p. 240; Huygens, 1888, vol. 2, p. 305.

4 Huygens, 1888, vol. 1, p. 401; Huygens, 1888, vol. 2, p. 306; Huygens, 1888, vol. 15, pp. 169, 254, 255; Van Helden, 1968, p. 225.

5 Wren to Neile, 1 October 1661, Royal Society MS EL.W.3 no. 2. Wren probably describes his own model in Van Helden, 1968, p. 226; see Figure 1.

6 See Van Helden, 1968, pp. 223, 224–5; Wren to Neile, 1 October 1661, Royal Society MS EL.W.3 no. 2.

7 See Van Helden, 1968, p. 217; Wren to Neile, 1 October 1661, Royal Society MS EL.W.3 no. 2. One of Wren's copies of *Systema Saturnium* is in the Bodleian Library, shelfmark Savile K.18; note Huygens, 1888, vol. 3, p. 453.

8 See Huygens, 1888, vol. 3, p. 368; T. Birch, 1756, vol. 1, p. 41.

9 Wren to Neile, 1 October 1661, Royal Society MS EL.W.3 no.2. Moray wrote to Huygens of their attempts to persuade Wren to produce his treatise: 'nous eusmes tous de la peine à luy faire accorder', Huygens, 1888, vol. 3, p. 425.

10 Huygens, 1888, vol. 3, pp. 368, 404, 415–24, 425–6; T. Birch, 1756, vol. 1, p. 66.

11 For the subsequent discussion, see Huygens, 1888, vol. 3, pp. 386, 399, 404, 408, 413, 437; Huygens, 1888, vol. 4, pp. 7, 24, 27, 34, 40–4, 62, 83, 145–6, 150–1, 155; T. Birch, 1756, vol. 1, p. 68.

12 See Huygens, 1888, vol. 3, pp. 368, 415–18, 425–6; Huygens, 1888, vol, 4, p. 83.

13 See Sprat, 1667, pp. 314–15; note also Wren, 1750, p. 242.

14 There is no reference to Wren's theory in Birch's *History*. For the only direct reference to Wren's ideas, see H. W. Robinson & W. Adams, 1935, p. 206.

15 Crossley, 1847, pp. 126–7; T. Birch, 1772, vol. 4, p. 118; T. Birch, 1756, vol. 1, p. 505.

16 For parallel work of other English scientists, see T. Birch, 1772, vol. 4, pp. 424, 487, 498–9; Waller, 1705, pp. 127–8; William Croone at British Library MS Add. 6193, fo. 108; Power, 1664, pp. 78–81.

17 For reports on Neile's telescopes, see Derham, 1726, p. 260; Huygens, 1888, vol. 15, pp. 70–1; Monconys, 1665, vol. 2, p. 77.

18 Van Helden, 1968, p. 221; Derham, 1726, p. 260; C. Wren, 1750, p. 205.

19 C. Wren, 1750, p. 198; T. Birch, 1756, vol. 1, p. 20; Huygens, 1888, vol. 22, p. 573.

20 There is reason to believe that it is correct, since Wren objected to aspects of Wallis's account that implied Wallis's priority regarding the demonstration, but apparently not the the story of its genesis, see A. R. Hall & M. B. Hall, 1965, vol. 6, pp. 280–1.

21 A. R. Hall & M. B. Hall, 1965, vol. 5, pp. 390–1. The earliest direct reference to Wren's demonstration and application is in a letter from John Collins to James Gregory, 2 February 1668/9, H. W. Turnbull, 1939, pp. 65–6.

22 T. Birch, 1756, vol. 2, p. 377. The term 'model' could refer to a design on paper, but it seems that this was an actual model, see *ibid.*, p. 399.

23 T. Birch, 1756, vol. 2, p. 379. The demonstration and corollary were registered the same day, Royal Society MS RB., iv, 71–2.

24 See *Philosophical Transactions*, 21 June 1669, vol. 4, no. 48, pp. 961–2; *ibid.*, 15 November 1669, vol. 4, no. 53, pp. 1059–60. For Wren's demonstration, see Whiteside, 1960, and Huxley, 1960.

25 See A. R. Hall & M. B. Hall, 1965, vol. 6, pp. 163, 306; T. Birch, 1756, vol. 2, p. 399.

26 See T. Birch, 1756, vol. 2, p. 379: 'any irregularity, made by the encountering of one another being immediately rectified'.

27 See T. Birch, 1756, vol. 2, pp. 377, 379; A. R. Hall & M. B. Hall, 1965, vol. 6, pp. 425–6.

28 See T. Birch, 1756, vol. 2, pp. 379, 382, 388, 399, 416. For Hooke's engine, see *ibid.*, p. 463.

29 See C. Wren, 1750, p. 198; Wren to Petty, 1656, see Bennett, 1973, p. 147 ('we . . . discover exactly her [the Moon's] various Inclinations, and herein Hevelius's Errors'); Sprat, 1667, p. 315. One of Wren's copies of Hevelius's *Epistolae II. Prior de motu lunae libratorio . . .* (Danzig, 1654) is in the Bodleian Library, shelfmark Savile B.15(2); another is listed in the sale catalogue of his library, see Bennett, 1974, p. 468.

30 For Wren's copies, see Bodleian Library, Savile B.15(1), and Bennett, 1974, p. 468.

31 C. Wren, 1750, p. 205; compare *ibid.*, pp. 200, 205, with Wilkins, 1802, vol. 1, pp. 26, 47–9.

32 Bennett, 1973, p. 147; note also Sprat, 1667, p. 315.

33 T. Birch, 1756, vol. 2, p. 139; note A. R. Hall & M. B. Hall, 1965, vol. 3, p. 297.

34 T. Birch, 1756, vol. 2, p. 139; note also *ibid.*, p. 140; *Philosophical Transactions*, 6 May 1667, vol. 2, no. 25, p. 459; *ibid.*, 11 November 1667, vol. 2, no. 29, pp. 541–4.

35 See King, 1955, pp. 94–6. Christopher Towneley was a patron of Horrox and Crabtree, and preserved their correspondence with Gascoigne, E. G. R. Taylor, 1954, pp. 81, 83, 216–17.

36 See E. G. R. Taylor, 1954, pp. 85, 229, 234, 357; G. H. Turnbull, 1952, pp. 114–15. Hooke included Ward along with Rooke and Wren as the foremost representatives of English observational astronomy, Hooke, 1665, preface.

37 See G. H. Turnbull, 1952, p. 112; M. Wren, 1659, preface.

38 See Huygens, 1888, vol. 3, p. 286; Neile and Moray to Wren, 17 May 1661, C. Wren, 1750, pp. 210–11. Perhaps Wren had already made a globe in connection with his 'Hypothesis of the Moon's Libration, in Solid', *ibid.*, p. 198.

39 See T. Birch, 1756, vol. 1, p. 33; C. Wren, 1750, p. 210.
40 See Huygens, 1888, vol. 3. p. 312; Moray to Wren, 13 August 1661, C. Wren, 1750, p. 211.
41 See Huygens, 1888, vol. 3, p. 317; A. R. Hall & M. B. Hall, 1965, vol. 1, p. 420. For the inscription see C. Wren, 1750, p. 211, here *Parentalia* can be relied on, since the globe was in Wren's son's possession, see J. Ward, 1740, p. 100; British Library MS Add. 25 071, fo. 97.
42 Sorbière, 1709, p. 34; note also Sprat, 1668, p. 151; Monconys, 1665, vol. 2, p. 82.
43 See Huygens, 1888, vol. 3, pp. 317, 320, 355; A. R. Hall & M. B. Hall, 1965, vol. 1, p. 420.
44 See Huygens 1888, vol. 7, p. 555; A. R. Hall & M. B. Hall, 1965, vol. 2, pp. 221, 307; T. Birch, 1756, vol. 1, pp. 468–9.
45 See T. Birch, 1756, vol. 2, pp. 156, 160. A lunar globe is included in a portrait of Wren by Verrio, Kneller and Thornhill, in the possession of Oxford University.
46 See Bennett, 1973, p. 147; C. Wren, 1750, p. 240.
47 See C. Wren, 1750, p. 198; Sprat, 1667, p. 315.
48 See Hooke, 1674, in Hooke, 1679, preface; Wallis in Halley, 1705, vol. 2, p. 293; Whiston, 1715, pp. 29–30.
49 See J. Ward, 1740, p. 104; note also David Gregory's note at Royal Society MS Gregory Volume, fo. 71. Ward's information came from Hodgson, and for Hodgson's links with Wren, see Baily, 1966, pp. 223, 238.
50 J. Pound, 'A rectification of the motions of the five satellites of Saturn', *Philosophical Transactions*, January–April 1718, vol. 30, no. 355, pp. 768–9; King, 1955, pp. 63–4.
51 See J. Ward, 1740, p. 105; Baily, 1966, p. 64; Flamsteed MS between pp. 244 and 245 of a copy of *Parentalia,* National Maritime Museum, 158a; Hodgson, 1706, preface.
52 See J. Ward, 1740, p. 105; Pound, 'A rectification of the motions of the five satellites of Saturn', *Philosophical Transactions*, January–April 1718, vol. 30, no. 355, p. 769.
53 See an annotation of his own copy of J. Ward, 1740, British Library, 611 m 16, p. 106; note also Sprat, 1667, p. 314; T. Birch, 1756, vol. 2, p. 313.
54 Sprat, 1667, p. 314; Monconys, 1665, vol. 2, p. 73.
55 Waller, 1705, p. 503; note also Derham, 1726, p. 272.

Chapter 5

1 See, for example, T. Birch, 1756, vol. 1, pp. 124, 131, 183–92; Astel, 1767, pp. 91, 108, 118, 125.
2 See Huygens, 1888, vol. 4, p. 296; Sprat, 1667, p. 248; Monconys, 1665, vol. 2, p. 27; Astel, 1767, p. 97; Wren at Royal Society MS EL.W.3 no. 3.
3 H. W. Robinson & W. Adams, 1935, pp. 188, 218, 375; C. Wren, 1750, p. 240; Tanner, 1926, vol. 2, p. 115.
4 See Grew, 1681, p. 364; note also *Biographia Britannica*, vol. 6, part 2, p. 4363.

5 See Sprat, 1667, p. 316; note also H. W. Robinson & W. Adams, 1935, pp. 220, 284, 288; C. Wren, 1750, p. 240.

6 See C. Wren, 1750, p. 240; note also T. Birch, 1756, vol. 3, pp. 395–6.

7 C. Wren, 1750, p. 240. For his father's interest, see a note in his copy of *Pseudodoxia epidemica,* Bodleian Library, 0.2.26 Art. Seld., p. 312.

8 C. Wren, 1750, p. 206; note also T. Birch, 1756, vol. 2, pp. 49, 51, 54, 57, 59; Wren to Brouncker, 30 July 1663, Royal Society MS EL.W.3. no. 3 (compare C. Wren, 1750, p. 199).

9 T. Birch, 1756, vol. 2, p. 440. For a later reference, see Royal Society Journal Book, vol. 9, p. 153.

10 See Brewster, 1855, vol. 2, p. 263 and Brewster, 1859.

11 See Sprat, 1667, pp. 313–14; note L. D. Patterson, 1952.

12 A. R. Hall & M. B. Hall, 1965, vol. 3, pp. 84–5; H. W. Robinson & W. Adams, 1935, pp. 162, 403, 412; note also *ibid.,* pp. 151, 161. Note Wren's interest T. Birch, 1756, vol. 1, p. 76; T. Birch, 1756, vol. 2, pp. 361, 435.

13 See T. Birch, 1756, vol. 2, pp. 146, 150, 151; Hooke, 1679, p. 196; note also H. W. Robinson & W. Adams, 1935, p. 272.

14 See Sprat, 1667, p. 314; Wilkins, 1668, p. 191; Derham, 1726, p. 390.

15 See T. Birch, 1756, vol. 1, pp. 70, 75, 500, 505; note also *ibid.,* pp. 495, 505–7; Huygens, 1888, vol. 5, pp. 149, 170–2; Huygens, 1888, vol. 16, pp. 354–5; T. Birch, 1772, vol. 6, pp. 500–1.

16 See T. Birch, 1756, vol. 2, p. 398; see also *ibid.,* p. 400; note also H. W. Robinson & W. Adams, 1935, p. 44; Hooke, 1679, p. 113.

17 C. Wren, 1750, p. 243. A marginal note at British Library MS Add. 25 071, fo. 101, describes this tract as being 'of late Date'; the (not altogether reliable) catalogue, British Library MS Lansdowne 698, no. 4, dates it to 1700; note also T. Birch, 1756, vol. 4, p. 491.

18 British Library MS Add 25 071, fo. 43, verso; note also Birch, 1756, vol. 1, pp. 69, 106, 108, 216.

19 See Birch, 1756, vol. 1, pp. 220, 223, Hooke later produced a map of the Pleiades, *ibid.,* p. 296; Hooke, 1665, pp. 241–2; Derham, 1726, pp. 272–3.

20 See Sprat, 1667, pp. 183–9; Whiston, 1738, pp. 19–20.

21 Pope, 1697, p. 116; compare Sprat, 1667, pp. 189–90.

22 Sorbière, 1709, p. 34; Huygens, 1888, vol. 4, pp. 458–9 – Huygens is replying to a letter from Moray, which has not survived.

23 See Pope, 1697, p. 117; T. Birch, 1756, vol. 1, p. 216.

24 See Whiston, 1715, p. 182; Whiston, 1738, 'An historical preface', p. 3; Waller, 1705, pp. 512, 514–17.

25 J. Ward, 1740, p. 109. For Ward's contacts with Wren's son see Bennett, 1973, pp. 135–8. According to *Biographia Britannica,* vol. 6, part 2, p. 4375, these papers came into the possession of William Jones. His library passed to the Earl of Macclesfield, but Wren's manuscripts cannot now be traced in the Macclesfield collection.

Chapter 6

1 See L. D. Patterson, 1949–50; Lohne, 1960; Koyré, 1965, pp. 180–4, 221–60; Westfall, 1967; Westfall, 1971, pp. 425–9; Whiteside, 1964 and 1970.

2 See Gilbert, 1958, Book 6; Bennett, 1981.

3 See Russell, 1964; Applebaum, 1969.

4 C. Wren, 1750, p. 204. Collatinus was associated with the overthrow of Tarquinius Superbus in 509 B.C.

5 See S. Ward, 1654, p. 29. Further details on Wren's ideas are in Bennett, 1975.

6 Wallis, 1659, pp. 80 (1st pagination)–73 (2nd pagination); note Whiteside, 1960, pp. 108–9.

7 See Koyré, 1965, pp. 201–20; McMullin, 1967, pp. 207–31; Drake, 1973, pp. 174–91.

8 Wren also had very relevant discussions with Huygens, probably in 1661, particularly those related to Huygens's application of the conical pendulum to a clock, see Bennett, 1975, p. 39.

9 R. Hooke, *An attempt to prove the motion of the earth from observations* (London, 1674), included in Hooke, 1679, which is reprinted in Gunther, 1920–67, vol. 8, see pp. 27–8.

10 See T. Birch, 1756, vol. 1, pp. 386, 395, 412–13, 414, 422, 456, 470, 473; T. Birch, 1756, vol. 2, p. 48; Plummer, 1940.

11 A. R. Hall & M. B. Hall, 1965, vol. 2, pp. 162–4, 177; Huygens, 1888, vol. 5, pp. 73, 79.

12 See T. Birch, 1756, vol. 1, p. 456; A. R. Hall & M. B. Hall, 1965, vol. 2, pp. 209, 213, 231–2.

13 See T. Birch, 1772, vol. 6, p. 501; T. Birch, 1756, vol. 1, pp. 504–5.

14 See T. Birch, 1756, vol. 2, p. 11; note A. R. Hall & M. B. Hall, 1965, vol. 2, pp. 341–2, 356 note 1.

15 A. R. Hall & M. B. Hall, 1965, vol. 2, pp. 353–4; compare Wallis, 1678, pp. 310–11, 321; Gunther, 1920, vol. 8, pp. 251–2.

16 Wallis's solution is preserved in manuscript at Bodleian MS Don. d. 45, fo. 283 (verso), headed: 'Problema. Dr Christopheri Wren, mihi propositur, 1st Jan. a.A° 1665.'

17 See A. R. Hall & M. B. Hall, 1965, vol. 2, pp. 353–4; T. Birch, 1756, vol. 2, p. 11.

18 See Gunther, 1920, vol. 8, pp. 215, 256–9; Bolton & Hendry, 1935, vol. 12, plate XLVII (reproduced from the collection of drawings by Wren at All Soul's College, Oxford, vol. 1, no. 3); note H. W. Robinson & W. Adams, 1935, pp. 120, 233.

19 See T. Birch, 1756, vol. 2, pp. 12, 13; Huygens, 1888, vol. 5, pp. 235–6, 262.

20 See C. Wren, 1750, p. 240; note also British Library MS Add. 25 071, fo. 91.

21 See T. Birch, 1756, vol. 2, p. 24; *Philosophical Transactions,* 1665, **1**, 17; A. R. Hall & M. B. Hall, 1965, vol. 2, pp. 359–67.

22 See T. Birch, 1756, vol. 2, pp. 54–9, 66, 74; T. Birch, 1772, vol. 6, p. 506.
23 See H. W. Robinson & W. Adams, 1935, pp. 307, 314, 427–30, 436; H. W. Turnbull *et al.,* 1959, vol. 2, p. 442.

Chapter 7

1 A. R. Hall & M. B. Hall, 1965, vol. 2, p. 561. For a slightly fuller account by Moray, see *ibid.,* p. 624; for Huygens, see *ibid.,* vol. 5, pp. 126–7; Huygens, 1888, vol. 6, pp. 383, 386; Huygens, 1888, vol. 16, p. 204; for Wallis, see A. R. Hall & M. B. Hall, 1965, vol. 5, p. 193.
2 Wren's original paper is at Royal Society MS CP.III(1) 43 (with copies at RB., iv, 29, and Boyle Papers, xx, fo. 157). It was printed in *Philosophical Transactions,* 1668–9, **3**, 867–8. For a translation see A. R. Hall & M. B. Hall, 1965, vol. 5, pp. 320–1. The theory is discussed in A. R. Hall, 1966, pp. 30–2, and in Westfall, 1971, pp. 203–6; note T. Birch, 1756, vol. 2, p. 315 (also *ibid.,* p. 140); A. R. Hall & M. B. Hall, 1965, vol. 5, pp. 117–18.
3 T. Birch, 1756, vol. 2, p. 335; note also *ibid.,* pp. 315, 337; A. R. Hall & M. B. Hall, 1965, vol. 5, pp. 117–18, 125, 134–5.
4 See Sprat, 1667, p. 312. For a reference by Newton, see Newton, 1947, p. 22.
5 See T. Birch, 1756, vol. 1, p. 68; note Huygens, 1888, vol. 5, p. 115.
6 See C. Wren, 1750, p. 183; compare British Museum MS Add. 25 071, fo. 36.
7 Note Hooke, 1665, preface; T. Birch, 1756, vol. 1, p. 266; G. H. Turnbull, 1952, p. 115.
8 Wilkins, 1649, p. 49. In the *Mathematical magic,* whose influence on Wren can be seen at several points, Wilkins wrote that: 'We see what strange discoveries of extreme minute bodies, (as lice, wheal-worms, mites, and the like) are made by the microscope' (Wilkins, 1802, vol. 2, p. 152).
9 See T. Birch, 1756, vol. 1, p. 21; C. Wren, 1750, pp. 210–11; Hooke, 1665, preface.
10 See T. Birch, 1756, vol. 1, pp. 21, 33; C. Wren, 1750, pp. 210–11.
11 See Power, 1664, p. 82; Hooke, 1665, preface.
12 C. Wren, 1750, pp. 204–5; for his idea that chemistry could aid mechanical hypotheses, see *ibid.,* p. 221.
13 See Monconys, 1665, vol. 2, p. 74; note Calvert, 1936, pp. 50–2.
14 See Monconys, 1665, vol. 2, pp. 74–5. Oldenburg later sent Monconys a perspectograph, see *ibid.,* pp. 63–7; A. R. Hall & M. B. Hall, 1965, vol. 5, pp. 285–91.
15 Latham & Matthews, 1972, vol. 8, pp. 310, 318; note also *ibid.,* p. 229.
16 See Grew, 1681, p. 376. Some drawings relating to the perspectograph are in the collection at All Souls College, Oxford, vol. 4, p. 156.
17 T. Birch, 1756, vol. 2, pp. 132, 133, 154, 157. Wren's account is at Royal Society MS CP.II.1; there is a copy at British Library MS Sloane 3323, fo. 283 (verso).

Chapter 8

1 See Ch. 3. H. W. Robinson & W. Adams, 1935, shows that they maintained their early contacts.

2 Bennett, 1973, p. 147. For other references, see Sprat, 1667, p. 317; C. Wren, 1750, p. 199; *Philosophical Transactions,* 1665–6, **1**, 128–30, 266; *Philosophical Transactions,* 1668, **3**, 672–82; A. R. Hall & M. B. Hall, 1965, vol. 2, pp. 334–8, 616; A. R. Hall & M. B. Hall, 1965, vol. 4, pp. 9, 55, 350–69; A. R. Hall & M. B. Hall, 1965, vol. 7, pp. 561–3. Hooke, 1665, p. 144; Huygens, 1888, vol. 5, p. 212; Evelyn, 1697, p. 166.

3 See C. Wren, 1750, p. 222. For comments on generation, see T. Birch, 1756, vol. 3, pp. 347, 350; Bolton & Hendry, 1923, vol. 19, p. 140.

4 See Sprat, 1667, p. 317; C. Wren, 1750, p. 243; T. Birch, 1756, vol. 1, pp. 233, 234, 269; Monconys, 1665, vol. 2, pp. 53–4.

5 See Webster, 1971. For a reference by Dean Wren to Scarburgh's splenectomy experiment, see a note in his copy of Bacon's *Sylva sylvarum,* Bodleian Library T.11.20. Th., *New Atlantis,* p. 35.

6 For a detailed discussion of this whole question, see Bennett, 1976, pp. 59–62.

7 Goodall, 1684, 'An historical account of the College's proceedings against empiricks and unlicensed practisers', 'The Epistle Dedicatory'.

8 C. Wren, 1750, p. 221. For the dating, see Bennett, 1976, p. 63, note 30.

9 See T. Birch, 1756, vol. 2, pp. 22–3, 25–6, 27, 29, 31; Huygens, 1888, vol. 5, p. 320. Boyle later said that he had frequently done experiments of this kind, T. Birch, 1756, vol. 3, p. 84.

10 T. Birch, 1756, vol. 2, p. 22. For Wilkins on experiments to demonstrate the mechanical force generated by inflating a bladder, see T. Birch, 1756, vol. 1, p. 36.

11 Willis, 1684, *The anatomy of the brain,* p. 105; see also pp. 105, 111. Note Hierons & Meyer, 1964.

12 See T. S. Patterson, 1931; Guerlac, 1953 and 1954; Partington, 1956; Debus, 1964; McKie in Underwood, 1953, vol. 1, pp. 469–88; Dewhurst, 1963, pp. 12–15.

13 Sprat, 1667, p. 316; note also *ibid.,* p. 218.

14 The letter is quoted from a copy by Oldenburg since this source is the closest available to the original. Where Oldenburg reads 'if' here, both C. Wren Jnr (1750, p. 226) and Abraham Hill (British Museum MS Sloane 2903, fo. 105) have read 'that'.

15 H. W. Robinson & W. Adams, 1935, p. 344. On 6 February 1689/90 Wren and Hooke discussed 'theory of Niter air flame', see Gunther, 1920, vol. 10, p. 185.

16 T. Birch, 1756, vol. 3, p. 403; note H. W. Robinson & W. Adams, 1935, p. 355.

17 See Guerlac, 1953, p. 339; note also Boyle in T. Birch, 1756, vol. 2, p. 287.

18 C. Wren, 1750, pp. 222–3; compare Sprat, 1667, pp. 312–13.

19 See Waller, 1705, pp. 7–8; T. Birch, 1772, vol. 1, p. 41; T. Birch, 1756,

vol. 3, p. 464; Derham, 1726, p. 1; Middleton, 1965, p. 145.

20 See T. Birch, 1756, vol. 1, pp. 5, 12, 77; T. Birch, 1772, vol. 1, p. 34; T. Birch, 1772, vol. 3, p. 208; A. R. Hall & M. B. Hall, 1965, vol. 4, p. 186.

21 See Sprat, 1667, p. 313; Middleton, 1964, pp. 102, 374.

22 Monconys, 1665, vol. 2, p. 56. For other contemporary references to balances by Wren, see Frank, 1972, p. 215; H. W. Robinson & W. Adams, 1935, pp. 372–3.

23 See G. H. Turnbull, 1952, pp. 111, 123; Vaughan, 1838, vol. 2, p. 464; Monconys, 1665, vol. 2, p. 53; Sprat, 1667, p. 313.

24 See Sprat, 1667, p. 313; Grew, 1681, p. 358; T. Birch, 1756, vol. 1, p. 74; T. Birch, 1756, vol. 3, p. 222; Biswas, 1967; G. J. Symons, 1891.

25 T. Birch, 1772, vol. 3, p. 142; note also T. Birch, 1772, vol. 2, p. 491,

26 Included in the 'Heirloom' copy of *Parentalia,* and printed in the 1965 Gregg Press reprint.

27 See Monconys, 1665, vol. 2, p. 56; note also *ibid.,* p. 74.

28 T. Birch, 1756, vol. 1, p. 271; note also *ibid.,* pp. 313, 315.

29 Royal Society MS EL.W.3 no. 3. For other references see C. Wren, 1750, p. 198; Waller, 1705, p. 554; A. R. Hall & M. B. Hall, 1965, vol. 4, p. 372.

30 See T. Birch, 1756, vol. 1, p. 341; note *ibid.,* pp. 300, 304, 305, 337, 344. For other contemporary references, see Sprat, 1667, pp. 312–13; Astel, 1767, pp. 124–5; Grew, 1681, pp. 357–8.

31 See T. Birch, 1756, vol. 1, p. 68; note also *ibid.,* p. 74, and a letter from William Croone to Henry Power, British Library MS Add. 6193, p. 106.

Chapter 9

1 T. Birch, 1756, vol. 1, p. 230; note also A. R. Hall, & M. B. Hall, 1965, vol. 2, pp. 44–5; Astel, 1767, p. 110; Clark, 1891, vol. 4, pp. 71–2.

2 T. Birch, 1756, vol. 2, p. 115; Waller, 1705, pp. 12–13.

3 See A. R. Hall & M. B. Hall, 1965, vol. 4, p. 204; Downes, 1968, p. 33; C. Wren, 1750, opposite p. 276; Strauss, 1954, p. 226.

4 See T. Birch, 1756, vol. 2, pp. 117, 118, 461, 464, 465; T. Birch, 1756, vol. 4, pp. 142, 146–9, 152; Sprat, 1667, p. 191; *Philosophical Transactions,* 1673, **8**, 6010–15; T. Birch, 1772, vol. 6, p. 495. vol. 6, p. 495.

5 C. Wren, 1750, pp. 196–7; compare Hooke, 1665, preface.

6 See Smith, 1704, p. 14; Clark, 1813, vol. 4, p. 704; Rigaud 1841, vol. 2, p. 217; T. Birch, 1772, vol. 6, p. 586.

7 See A. R. Hall & M. B. Hall, 1965, vol. 6, pp. 52–4, 447–8, 458–60; A. R. Hall & M. B. Hall, 1965, vol. 7, pp. 395–7.

8 The trip was being planned at least as early as March 1665 (Huygens, 1888, vol. 5, p. 286), and Wren left in early July; note Wheatley, 1960, vol. 3, pp. 304–6; Bolton & Hendry, 1923, vol. 5, p. 14; A. R. Hall & M. B. Hall, 1965, vol. 2, pp. 428–9. For accounts of the visit, see Bolton & Hendry, 1923, vol. 18, pp. 177–80; Sekler, 1956, pp. 44–50; Whinney, 1958.

9 See A. R. Hall & M. B. Hall, 1965, vol. 2, pp. 428–9, 480–2, 517–19; A. R. Hall & M. B. Hall, 1965, vol. 3, pp. 11–12, 36–8, 48, 81–5; C. Wren, 1750, p. 261. For the French circle see Brown, 1934.

10 See A. R. Hall & M. B. Hall, 1965, vol. 2, pp. 480–2, and compare C. Wren. 1750, pp. 261–2.

11 See Huygens, 1888, vol. 5, pp. 212, 228, 235, 249, 262, 266.

12 See A. R. Hall & M. B. Hall, 1965, vol. 2, pp. 480, 519; A. R. Hall & M. B. Hall, 1965, vol. 3, pp. 38, 83–5.

13 See Bolton & Hendry, vol. 10, pp. 14, 45–53; Colvin, 1954.

14 See C. Wren, 1750; note Ch.1.

15 The body of St Michael Cornhill (1669) has sometimes been attributed to Wren but the vestry minutes indicate that the design is not his, see Bolton & Hendry, 1923, vol. 19, pp. 45–6.

16 Apart from the surviving buildings, the chief specific sources on the churches are Bolton & Hendry, 1923, vols. 9, 10, 19; Summerson, 1952, 1970; Clayton, 1848; G. H. Birch, 1819; Clarke, 1819; Daniell, 1895; Bumpus, 1908; Cobb, 1948; Davison, 1923; Macmurdo, 1883; Reynolds, 1922; A. F. Taylor, 1881; Davis, 1935. Among general works on Wren, see Summerson, 1953; Briggs, 1953; Fuerst, 1956; Sekler, 1956; Whinney, 1971; Downes, 1971.

17 Fuerst, 1956, p. 9, seems to be wrong here in suggesting that Wren himself divided the site to provide a square interior, for see G. H. Birch, 1819, p. 41; A. F. Taylor, 1881, p. 33; Daniell, 1895, p. 238, and compare the print in the 'Heirloom' copy of *Parentalia*.

18 C. Wren, 1750, pp. 305–6; for the original, see Bolton & Hendry, 1923, vol. 11, p. 22.

19 For the date, see Fuerst, 1956, pp. 20, 224; Bolton & Hendry, 1923, vol. 19, p. 11.

20 Galleries were added at St Mary-le-Bow and Briggs, in 1953, p. 125, attributes them to Wren. However, McCulloch, 1964, p. 15, says they were eighteenth-century additions, removed in 1867. There is insufficient evidence (Bolton & Hendry, 1923, vol. 19, pp. 11–12) for the claim that Wren provided side galleries at St Bride, only under pressure from the parish.

21 For dating, see Bolton & Hendry, 1923, vol. 10, p. 52; Bolton & Hendry, 1923, vol. 19, p. 49

22 For dating, see Fuerst, 1956, p. 224; Bolton & Hendry, 1923, vol. 19, pp. 21–2.

23 C. Wren, 1750, p. 201. David Gregory wrote in 1698,

Mr C. Wren says that he is in possession of a method of explaining gravity mechanically. He smiles at Mr Newton's belief that it does not occur by mechanical means, but was introduced originally by the Creator
H. W. Turnbull, *et al* (1959), vol. 4, pp. 226–7.

Chapter 10

1 See A. R. Hall & M. B. Hall, 1965, vol. 5, pp. 263, 344, 347, 360–2, 363; A. R. Hall & M. B. Hall, 1965, vol. 6, p. 270; A. R. Hall & M. B. Hall, 1965, vol. 7, pp. 65–6, 162–6; Loemker, 1956, vol. 2, p. 719; note also Keill, 1720, p. 194.
2 C. Wren, 1750, p. 351. It is clear from the following paragraph that Wren is using the term 'orders' here in a very general sense.
3 C. Wren, 1750, p. 353. This has previously been seen as contradicting Wren's idea of natural beauty, Fuerst, 1956, note 1023.
4 Perrault, 1683 (the Royal Society were given notice of the publication, see T. Birch, 1756, vol. 4, p. 205); Perrault, 1708.
5 See his discussion of proportions as loose conventions based on visual judgments, C. Wren, 1750, p. 353.
6 Fréart, 1650; Fréart, 1664.

References

A list of full references to works cited in abbreviated form in the notes

Allen, P. (1949). Scientific studies in the English universities of the seventeenth century. *Journal of the History of Ideas*, **10**, 219–53.

Applebaum, W. (1969). *Kepler in England: the reception of Keplerian astronomy in England, 1599–1687*. Ph. D. Dissertation. University Microfilms Inc., Michigan.

Armitage, A. (1950). 'Borell's hypothesis' and the rise of celestial mechanics. *Annals of Science*, **6**, 268–82.

– (1960). 'William Ball, F.R.S. (1627–1690). *Notes and Records of the Royal Society of London*, **15**, 167–72.

Astel, F. ed. (1767). *Familiar letters of Abraham Hill*. London.

Baily, F. ed. (1966). *An account of the Rev. John Flamsteed*. London.

Beer, E.S. de. ed. (1955). *The diary of John Evelyn*. Oxford.

Bennett, J.A. (1973). A study of *Parentalia*, with two unpublished letters of Sir Christopher Wren. *Annals of Science*, **30**, 129–47.

– (1974). *Studies in the life and work of Sir Christopher Wren*. Ph. D. Dissertation. University of Cambridge.

– (1975). Hooke and Wren and the system of the world. *British Journal for the History of Science*, **8**, 32–61.

– (1976). A note on theories of respiration and muscular action in England c. 1660. *Medical History*, **20**, 59–69.

– (1981). Cosmology and the magnetical philosophy, 1640–1680. *Journal for the History of Astronomy*, **12**, 165–77.

Billingsley, H. trans. (1570). *The elements of geometrie . . . of Euclide*. London.

Biographia Britannica (1776): *or, the lives of the most eminent persons who have flourished in Great Britain and Ireland . . .*, vol. 6. London.

Birch, G.H. (1819). *London churches of the seventeenth and eighteenth centuries*. London.

Birch, T. (1756–7). *A history of the Royal Society of London for improving of natural knowledge, from its first rise . . .*, 4 vols. London.

– ed. (1772). *The life and works of the Honourable Robert Boyle*, 2nd edn, 6 vols. London.

Biswas, A.K. (1967). The automatic rain-gauge of Sir Christopher Wren, F.R.S. *Notes and Records of the Royal Society of London*, **22**, 94–104.

Bliss, P. ed. (1813–20). *Athenae Oxonienses*, 3rd edn, 4 vols. London.

Boas, M. (1962). *The scientific renaissance 1450–1630*. London.
Bolton, A.T. & H.D. Hendry, eds. (1923–43). *Wren Society*, 20 vols. London.
Brewster, D. (1855). *Memoirs of the life, writings, and discoveries of Sir Isaac Newton*, 2 vols. Edinburgh.
– (1859). On Sir Christopher Wren's cypher, containing three methods of finding the longitude. *Report of the British Association*, notes and abstracts p. 34.
Briggs, W.S. (1953). *Wren the incomparable*. London.
Brown, H. (1934). *Scientific organizations in seventeenth century France*. Baltimore.
Bumpus, T.F. (1908). *London churches ancient and modern*. London.
Calvert, H.R. (1936–7). Thompson of Hosier Lane, instrument makers of the seventeenth century. *Isis*, **26**, 50–2.
Carlyle, T. ed. (1904). *The letters and speeches of Oliver Cromwell*. London.
Caröe. W.D. (1923). *'Tom Tower' Christ Church, Oxford . . . .* Oxford.
Clark, A. ed. (1891–1900). *The life and times of Anthony Wood . . . described by himself,* 5 vols. Oxford.
– ed. (1898). *'Brief Lives', chiefly of contemporaries . . . ,* 2 vols. Oxford.
Clarke, C. (1819). *Architectura ecclesiastica Londini*. London.
Clayton, J. (1848–9). *The dimensions, plans, elevations, and sections of the parochial churches of Sir Christopher Wren . . . .* London.
Cobb, G. (1948). *The old churches of London*. London.
Colie, R.L. (1960). Dean Wren's marginalia and early science at Oxford. *Bodleian Library Record,* **6**, 541–51.
Colvin, H.M. (1951). Roger North and Sir Christopher Wren. *Architectural Review,* **111**, 257–60.
– (1954). *A biographical dictionary of English architects 1666–1840*. London.
Copeman, W.S.C. (1960). Dr Jonathan Goddard, F.R.S. (1617–1675). *Notes and Records of the Royal Society of London,* **15**, 69–77.
Crossley, J. ed. (1847). *The diary and correspondence of Dr. John Worthington,* vol. 1, Manchester.
Cunningham, W. (1559). *The cosmographical glasse*. London.
Daniell, A.E. (1895). *London City Churches*. London.
Davis, E.J. (1935). The parish churches of the City of London. *Journal of the Royal Society of Arts,* **83**, 895–912.
Davison, T.R. (1923). *Wren's City Churches*. London.
Debus, A.G. (1964). The Paracelsian aerial niter. *Isis,* **55**, 43–61.
Derham, W. ed. (1726). *Philosophical experiments and observations of Dr. Robert Hooke*. London.
Descartes, R. (1965). *Discourse on method, optics, geometry, and meteorology,* trans. P.J. Olscamp. New York.
Dewhurst, K. (1963) *John Locke, physician and philosopher*. London.
Digges, L. (1556). *A booke named Tectonicon*. London.
Digges, L. & T. Digges. (1591). *A geometrical practical treatize named Pantometria*. London.
Downes, K. (1968). John Evelyn and architecture: a first inquiry, in J.

Summerson, ed., *Concerning architecture* . . . . London.
– (1971). *Christopher Wren*. London.
Drake, S. (1973). Galileo's 'Platonic' cosmogony and Kepler's Prodromus. *Journal for the History of Astronomy,* **4,** 174–91.
Evelyn, J. (1697). *Numismata. A discourse of medals, ancient and modern* . . . . London.
– (1706). *An account of architects and architecture*. London.
Feindel, W. ed. (1965). *The anatomy of the brain and nerves. Tercentenary edition,* 2 vols. Montreal.
Flamsteed, J. (1680). *The doctrine of the sphere* . . . . London. (In J. Moore, 1681*a, A new systeme of the mathematicks* . . . . London.)
Frank Jnr, R.G. (1972–3). John Aubrey, F.R.S., John Lydall, and science at Commonwealth Oxford. *Notes and Records of the Royal Society of London,* **27,** 193–217.
– (1973). Science, medicine and the universities of early modern England. *History of Science,* **11,** 194–216, 239–69.
Fréart, R. (1650). *Parallèle de l'architecture antique et de la moderne*. Paris.
– (Trans. J. Evelyn, 1664, *A parallel of the ancient architecture with the modern* . . . . London.)
Fuerst, V. (1956). *The architecture of Sir Christopher Wren*. London.
Gardiner, R.B. (1889). *The registers of Wadham College, Oxford,* part 1, 1613–1719. London.
Gellibrand, H. (1635). *A discourse mathematical on the variation of the magneticall needle* . . . . London.
Gilbert, W. (1958). *De magnete,* trans. P.F. Mottelay. New York.
Goodall, C. (1684). *The Royal College of Physicians* . . . . London.
Gould, R.T. (1960). *The marine chronometer*. London.
Grew, N. (1681). *Musaeum Regalis Societatis*. London.
Guerlac, H. (1953). John Mayow and the aerial nitre. *Actes du VIIe Congrès International d'Histoire des Sciences,* pp. 332–49.
– (1954). The poet's nitre. *Isis,* **45,** 243–55.
Gunter, E. (1662). *The works of Edmund Gunter,* 4th edn, London.
Gunther, R.T. (1920–67). *Early science in Oxford,* 15 vols. Oxford.
Hall, A.R. (1965). Wren's problem. *Notes and Records of the Royal Society of London,* **20,** 140–4.
– (1966–7). Mechanics and the Royal Society, 1668–70. *British Journal for the History of Science,* **3,** 24–38.
Hall, A.R. & M.B. Hall, eds. (1965–77). *The correspondence of Henry Oldenburg,* 11 vols. Madison and London.
Halley, E. (1705). *Miscellanea curiosa* . . . . London.
Hartlib, S. (1655). *The reformed commonwealth of bees* . . . . London.
Hearne, T. ed. (1725). *Peter Langtoft's chronicle*. Oxford.
Hierons, R. & A. Meyer. (1964). Willis's place in the history of muscle physiology. *Proceedings of the Royal Society of Medicine,* **57,** 687–92.
Hill, A. (1677). Some account of the life of Dr Isaac Barrow, in J. Tillotson, ed., 1683, *The works of Isaac Barrow,* vol. 1. London.
Hill, C. (1965). *Intellectual origins of the English Revolution*. Oxford.

Hiscock, W.G. ed. (1937). *David Gregory, Isaac Newton and their circle. Extracts from David Gregory's memoranda 1677–1708.* Oxford.
Hodgson, J. (1706). *The theory of navigation demonstrated . . . .* London.
Holder, W. (1678). *A supplement to the Philosophical Transactions of July, 1670.* London.
– (1694). *A discourse concerning time, with application of the natural day, and lunar month, and solar year, as natural . . . .* London.
Hooke, R. (1665). *Micrographia, or some physiological descriptions of minute bodies . . . .* London.
– (1679). *Lectiones Cutlerianae . . . .* London. (Reprinted in R. Gunther, 1920–67, *Early science in Oxford,* vol. 8. Oxford.)
Hutchison, H.F. (1976). *Sir Christopher Wren, a biography.* London.
Huxley, G.H. (1960). The geometrical work of Christopher Wren. *Scripta Mathematica,* **25,** 201–8.
Huygens, C. (1888–1950). *Oeuvres complètes,* 22 vols. The Hague.
Johnson, F.R. (1937). *Astronomical thought in Renaissance England.* Baltimore.
Josten, C.H. (1966). *Elias Ashmole (1617–1692),* 5 vols. Oxford.
Keevil, J.J. (1952). Sir Charles Scarburgh. *Annals of Science,* **8,** 113–21.
Keill, J. (1720). *An introduction to natural philosophy . . . .* London.
King, H.C. (1955). *The history of the telescope.* London.
Koyré, A. (1965). *Newtonian studies.* London.
Latham, R. & W. Matthews, eds. (1972). *The diary of Samuel Pepys.* London.
Lawrence, P.D. & A.G. Molland. (1970). David Gregory's Inaugural Lecture at Oxford. *Notes and Records of the Royal Society of London,* **25,** 143–78.
Lister, M. (1699). *A journey to Paris in the year 1698,* 3rd edn, London.
Loemker, L.E. ed. (1956). *Philosophical papers and letters* [of W. von Leibniz], 2 vols. Chicago.
Lohne, J. (1960–1). Hooke *versus* Newton. *Centaurus,* **7,** 6–52.
McCulloch, J. (1964). *The pictorial history of St Mary-le-Bow.* London.
McKie, D. (1960). The origins and foundation of the Royal Society of London. *Notes and Records of the Royal Society of London,* **15,** 1–37.
McMullin, E. ed. (1967). *Galileo, man of science.* New York.
MacMurdo, A.H. (1883). *Wren's City churches.* London.
Maddison, R.E.W. (1951–2). Studies in the life of Robert Boyle, F.R.S., part 1, Robert Boyle and some of his foreign visitors. *Notes and Records of the Royal Society of London,* **9,** 1–35.
Middleton, W.E.K. (1964). *The history of the barometer.* Baltimore.
– (1965) A footnote to the history of the barometer. *Notes and Records of the Royal Society of London,* **20,** 145–51.
Milman, L. (1908). *Sir Christopher Wren.* London.
Monconys, B. de (1665–6). *Journal des voyages de Monsieur de Monconys . . . publié par le Sieur de Liergues son fils,* 2 vols. Lyon.
Moore, J. (1681a). *A new systeme of the mathematicks . . . .* London.
– (1681b). *A mathematical compendium; or, useful practices in arithmetick, geometry, etc. . . . ,* 2nd edn. London.

Munk, W. (1878). *The roll of the Royal College of Physicians of London,* 2nd edn, 3 vols. London.

Newton, I. (1947). *Mathematical principles of natural philosophy,* trans. A. Motte, revised F. Cajori. California.

Oughtred, W. (1652). *Clavis mathematicae denvo limata, sive pontius fabricata . . . Editio tertia auctior & ementatior. . . .* Oxford.

Partington, J.R. (1956). The life and work of John Mayow, (1641–1679). *Isis,* **47**, 217–30, 405–17.

Pascal, B. (1659). *Lettres de Dettonville. . . .* Paris.

Patterson, L.D. (1949–50). Hooke's gravitational theory and its influence on Newton. *Isis,* **40**, 312–41; **41**, 32–45.

– (1952). Pendulums of Wren and Hooke. *Osiris,* **10**, 277–321.

Patterson, T.S. (1931). John Mayow in contemporary setting. *Isis,* **15**, 47–96, 504–46.

Perrault, C. (1683). *Ordonnance des cinq espèces de colonnes selon la méthode des anciens.* Paris.

– (1708). *A treatise of the five orders of columns in architecture,* trans. J. James. London.

Pevsner, N. (1968). *An outline of European architecture.* London.

Plummer, H.C. (1940–1). Jeremiah Horrox and his *Opera Posthuma. Notes and Records of the Royal Society of London,* **3**, 39–52.

Pope, W. (1697). *Life of Seth Ward . . .,* London.

Power, H. (1664). *Experimental philosophy.* London.

Purver, M. (1967). *The Royal Society: concept and creation.* London.

Recorde, R. (1551). *The pathway to knowledge.* London.

– (1556). *The castle of knowledge.* London.

Reynolds, H. (1922). *The churches of the City of London.* London.

Rigaud, S.J. (1841). *Correspondence of scientific men of the seventeenth century,* 2 vols. Oxford.

Robertson, J.D. (1931). *The evolution of clockwork.* London.

Robinson, H.W. (1949–50). An unpublished letter of Dr. Seth Ward relating to the early meetings of the Oxford Philosophical Society. *Notes and Records of the Royal Society of London,* **7**, 68–70.

Robinson, H.W. & W. Adams, eds. (1935). *The diary of Robert Hooke, 1672–1680.* London. (Note also R. Gunther, 1920–67, *Early science in Oxford,* vols. 7, 10. Oxford.)

Ronan, C.A. & H. Hartley. (1960). Paul Neile, F.R.S. (1613–1686). *Notes and Records of the Royal Society of London,* **15**, 159–65.

Russell, J.L. (1964–5). Kepler's laws of planetary motion: 1609–1666. *British Journal for the History of Science,* **2**, 1–24.

Sekler, E.F. (1956). *Wren and his place in European architecture.* London.

Shapiro, B.J. (1969). *John Wilkins 1614–1672. An intellectual biography.* Berkeley.

Smith, T. (1704). *Vita . . . Edwardi Bernardi . . .,* in *D. Roberti Huntingtoni . . . Epistolae. . . .* London.

Sorbière, S. (1709). *A voyage to England . . . .* London.

Sprat, T. (1667). *The history of the Royal Society of London.* London.

Sprat, T. (1668). *Observations on Monsieur de Sorbier's voyage into England. Written to Dr. Wren, Professor of Astronomy in Oxford* . . . . London.

Stimson, D. (1931). Dr. Wilkins and the Royal Society. *Journal of Modern History*, **3**, 539–63.

Strauss, E. (1954). *Sir William Petty. Portrait of a genius.* London.

Summerson, J. (1949). The mind of Wren, in *Heavenly mansions and other essays on architecture.* London.

– (1952). Drawings for the London City churches. *Journal of the Royal Institute of British Architects*, 3rd series, **59**, 126–9.

– (1953). *Sir Christopher Wren.* London.

– (1960). Sir Christopher Wren, P.R.S. (1632–1723). *Notes and Records of the Royal Society of London*, **15**, 99–105.

– (1970). Drawings of London churches in the Bute Collection: a catalogue. *Architectural History*, **13**, 30–42.

Symons, G.J. (1891). A contribution to the history of rain gauges. *Quarterly Journal of the Royal Meteorological Society*, **12**, 127–42.

Tanner, J.R. ed. (1926) *Samuel Pepys, private correspondence and miscellaneous papers, 1679–1703.* London.

Taylor, A.F. (1881). *The towers and steeples designed by Sir Christopher Wren.* London.

Taylor, E.G.R. (1954). *The mathematical practitioners of Tudor and Stuart England.* Cambridge.

Turnbull, G.H. (1952–3). Samuel Hartlib's influence on the early history of the Royal Society. *Notes and Records of the Royal Society of London*, **10**, 101–30.

Turnbull, H.W. ed. (1939). *James Gregory tercentenary memorial volume.* London.

Turnbull, H.W. et al. eds. (1959–77). *The correspondence of Isaac Newton*, 7 vols. Cambridge.

Underwood, E.A. ed. (1953). *Science, medicine and history.* Oxford.

Van Helden, A. (1968). Christopher Wren's *De Corpore Saturni. Notes and Records of the Royal Society of London*, **23**, 213–29.

Vaughan, R. ed. (1838). *The Protectorate of Oliver Cromwell*, 2 vols. London.

Waller, R. ed. (1705). *The posthumous works of Robert Hooke.* London.

Wallis, J. (1655). *Eclipsis solaris Oxonii visae anno . . . 1654.* Oxford.

– (1657). *Operum mathematicorum.* Oxford.

– (1659). *Tractatus duo. Prior, de cycloide* . . . . Oxford.

– (1665–6). An essay of Dr. John Wallis, exhibiting his hypothesis about the flux and reflux of the sea. *Philosophical Transactions*, **1**, 263–89.

– ed. (1678). *Opera posthuma* . . . [of Jeremiah Horrox]. London.

– (1685). *A treatise of algebra, both historical and practical* . . . . London.

Ward, J. (1740). *Lives of the professors of Gresham College.* London.

Ward, S. (1653). *In Ishmaelis Bullialdi Astronomiae Philolaicae inquisitio brevis.* Oxford.

– (1654). *Vindiciae academiarum* . . . , with preface by J. Wilkins. Oxford.

– (1656). *Astronomia geometrica* . . . . London.

Warton, T. (1761). *The life and literary remains of Ralph Bathurst . . . .* London.

Waters, D.W. (1958). *The art of navigation in England in Elizabethan and early Stuart times.* London.

Watkins, R. ed. (1651). *Newes from the dead . . . .* 1st and 2nd impressions. Oxford.

Webster, C. (1971). The Helmontian George Thomson and William Harvey: the revival and application of splenectomy to physiological research. *Medical History,* **15**, 154–67.

Westfall, R.S. (1966–7). Robert Hooke and the law of universal gravitation. *British Journal for the History of Sciences,* **3**, 245–61.

– (1971). *Force in Newton's physics.* London.

Wheatley, H.B. ed. (1893–9). *S. Pepys' Diary,* 10 vols. London.

– ed. (1960). *Diary and correspondence of John Evelyn,* 4 vols. London.

Whinney, M.D. (1958). Sir Christopher Wren's visit to Paris. *Gazette des Beaux-Arts,* 6th period, **51**, 229–42.

– (1971). *Wren.* London.

Whiston, W. (1715). *Astronomical lectures, read in the publick schools at Cambridge.* London.

– (1738). *The longitude discovered by the eclipses, occultations, and conjunctions of Jupiter's planets . . . .* London.

Whiteside, D.T. (1960). Wren the mathematician. *Notes and Records of the Royal Society of London,* **15**, 107–11.

– (1964–5). Newton's early thoughts on planetary motion: a fresh look. *British Journal for the History of Science,* **2**, 117–37.

– (1970). Before the *Principia*: the maturing of Newton's thoughts on dynamical astronomy, 1664–1684. *Journal for the History of Astronomy,* **1**, 5–19.

Wilkins, J. (1640). *The discovery of a new world,* 2nd edn. London.

– (1649). *A discourse concerning the beauty of providence in all the rugged passages of it.* London.

– (1668). *An essay towards a real character and a philosophical language.* London.

– (1802). *The mathematical and philosophical works of John Wilkins,* 2nd edn, 2 vols. London.

Willis, T. (1684). *Dr. Willis's practice of physick.* The anatomy of the brain. London.

Wilson, L.G. (1961). William Croone's theory of muscular contraction. *Notes and Records of the Royal Society of London,* **16**, 158–78.

Wren Jnr, C. (1750). *Parentalia: or, memoirs of the family of the Wrens . . . ..* London.

Wren, M. (1659). *Monarchy asserted: or, the state of monarchicall and popular government . . . .* Oxford.

Yates, F. (1969). *Theatre of the world.* London.

Index

Printed in the United States
By Bookmasters